Papa Topside

Papa Topside

The Sealab Chronicles of
Capt. George F. Bond, USN

EDITED BY HELEN A. SIITERI

NAVAL INSTITUTE PRESS
Annapolis, Maryland

Library of Congress Cataloging-in-Publication Data

Bond, George F. (George Foote), 1924–1983
    Papa Topside : the Sealab chronicles of Capt. George F. Bond / edited by Helen A. Siiteri.
        p.   cm.
    Includes index.
    ISBN 1-55750-795-3 (alk. paper)
    1. Bond, George F. (George Foote), 1924–1983.
2. Divers—United States—Biography. 3. United States.
Navy—Biography. 4. Saturation diving. I. Title.
VM980.B66A3 1993
33297—dc20   92-33297
                                                  CIP

Printed in the United States of America on acid-free paper ♾

9 8 7 6 5 4 3 2

First printing

Frontispiece: Captain George F. Bond checks the chamber pressure gauge on the U.S. Navy's first Personnel Transfer Capsule, which rests on its side on the deck of the support barge *YFNB-12*. Decompression of the four Sealab I aquanauts inside the chamber took about one day. (U.S. Navy photo)
    The four Navy aquanauts were Lester E. Anderson; Robert A. Barth; Sanders W. Manning; and Robert E. Thompson.

Contents

Acronyms

AFIP	Armed Forces Institute of Pathology
ATA	atmosphere absolute
BOQ	bachelor officers' quarters
BuMed	Bureau of Medicine and Surgery
BuShips	Bureau of Ships
CNO	Chief of Naval Operations
DDC	deck decompression chamber
DeKomRaum	decompression chamber (*Helgoland*)
DMO	diving medical officer
DSSP	Deep Submersible Systems Program
DSSRG	Deep Submersible Systems Review Group
EDU	Experimental Diving Unit
FISSH	First International Saturation Study of Herring and Hydroacoustics
MDL	Mine Defense Laboratory
NASA	National Aeronautics and Space Administration
NavShips	Naval Ship Systems Command
NCSL	Naval Coastal Systems Laboratory
NOAA	National Oceanographic and Atmospheric Administration
NOTS	Naval Ordnance Testing Station
NSRDL	Naval Ship Research and Development Laboratory

ONR	Office of Naval Research
PTC	personnel transfer capsule
SDC	submersible decompression chamber
SitRep	situation report
SMO	submarine medical officer
SubMedCen	Submarine Medical Center
TAD	temporary additional duty

Foreword

Capt. George F. Bond, MC, USN, who made his mark as one of the U.S. Navy's undersea pioneers and the "Father of Saturation Diving," was, first and foremost, a medical practitioner to the family of Navy divers. Members of his squadron sought his advice in connection with their most personal problems and those of their families. He considered his assignment as senior medical officer an opportunity to practice family medicine in the truest sense. Submariners and deep-sea divers recognized him as one of their own—a man who spoke their language and shared their lives.

I knew him as perhaps no other associate did. During my years with him as coinvestigator, I rarely went anywhere in the Navy diving community that I did not hear stories of how he had gone out of his way to provide medical attention to sailors or their families. He was never too busy to share our burdens.

Dr. Bond was a man of vision and foresight, fortified with a sprinkling of fantasy and a single-mindedness of purpose. As early as 1957, he postulated that human beings would one day inhabit the seas as free agents. The transition from fantasy to fact was not easy. While industry could take immediate advantage of new technology, we were delayed in our progress from one phase to the next by a constant requirement to justify our goals to higher authority. However deterred by that authority, Dr. Bond lived by the adage "They's more'n one way to skin a cat."

The fact that a medical doctor and his small team were embarked on a totally new concept of deep-sea diving was not taken lightly by the dyed-in-the-wool "salvage" diving community. A dichotomy developed; the old school was reluctant to accept that the rules of thumb governing the life of the surface-supported diver did not apply to the saturation diving technique. This led to direct competition for facilities, divers, and the most necessary evil of all, funding.

With the successful completion of Sealab I, the salvage diving industry began to recognize the true value of saturation techniques. Besides increasing the amount of meaningful work one could expect from a diving team, such techniques had a favorable ratio of work accomplished to cost incurred. With this added stimulus, many large diving companies began to develop operational saturation systems for open-water use.

By 1970, one of the major diving companies completed its first 1,000-feet-of-seawater simulated dive in its new state-of-the-art, deep-diving complex. During the press conference that followed, the leading Navy engineer asked, "Hey, Doc, how do you like a diving complex that was designed and built by real engineers rather than you medical guys?" The only answer I could come up with represented my true feelings: "It's easy when the road has been paved."

If it were not for the development of saturation diving, the offshore petroleum industry as we know it would not have been feasible. Saturation diving using a personnel transfer capsule and a deck decompression chamber opened up the oil fields of the continental shelf for exploration and exploitation. The National Oceanic and Atmospheric Administration (NOAA) has accomplished fisheries studies, oceanography programs, marine engineering projects, and marine biology research using saturation diving systems. NOAA has credited this diving technique with making it possible "for scientists to live and work on the ocean floor for virtually unlimited time, allowing a nine-hour day of research before returning to the habitat."

Through his work and perseverance, dedication and

leadership, Dr. Bond held the torch that lit the way for others to follow. I am proud to say Papa Topside was an advocate for all Navy divers, was my coworker, and above all, was my friend.

Walt Mazzone

Acknowledgments

As George F. Bond had the remarkable good sense to keep a journal throughout his life, I set out to edit the existing rough draft of his lifetime as he recorded it. The help and good will of many people, particularly Mr. and Mrs. George F. Bond, Jr., made this project possible.

Alex Rynecki of Sausalito, California, introduced me to the world of diving and marine salvage when he handed me the assignment to assist him in researching salvage in the U.S. Navy for the Naval Sea Systems Command. Our research was used as a starting point for *Mud, Muscle, and Miracles,* brought to publication in 1990 by the late Capt. C. A. Bartholomew, USN. That is where I paid my dues. Alex Rynecki recommended that I start research on *Papa Topside* by calling his friend John Quirk, consultant to the Naval Coastal Systems Laboratory in Panama City, Florida. John Quirk helped me set up interviews with the Sealab aquanauts who were then living in Panama City.

I especially thank aquanaut Bob Barth for reading this narrative, in which he has a prominent role, and for commenting, "Everyone will remember these events differently, but this is as close to the truth as anyone will probably get." Of Bond, he said, "He was my friend. I loved him and I always will."

I thank Bob Bornholdt, who recalled how he, and probably hundreds like him, "longed with all my heart to be part—any part—of the Sealab project," and of his disappointment when this did not happen. Years later, as president of the Institute of Diving and while commanding officer of the Naval Experimental Diving Unit, Bornholdt was instrumental in raising the sunken hulk of Sealab I offshore Panama City and moving the habitat to the Museum of Man in the Sea, where it now rests.

I thank others in Panama City who shared their recollections, including aquanauts Wally Jenkins and Tom James, then director of the International Diving Museum, who provided access to collected materials and most of the photographs in this book.

Walter Mazzone patiently helped with technical references to the physiology of diving, identified dozens of faces in photographs taken twenty-five years before, and prepared clear and direct answers to my questions.

Scott Carpenter reviewed the Sealab II chapter, providing corrections and the reassuring comment: "It was great fun to hear him [Bond] speak again."

I thank my friends Martha Casselman Spaulding, Dorothy DeRenzo, and Alex Rynecki, who faithfully read an early draft and suggested many constructive changes; the late Dr. Albert Behnke and his wife, Ruth, who provided a missing section of correspondence and an unforgettable afternoon of their personal recollections of George Bond, their friend; to Barbara Desiderati of the Undersea and Hyperbaric Medical Society, Inc. for her prompt responses to my queries; and to Dorothy Parkinson and Douglas Hough, director of the Museum of Man in the Sea, in Panama City Beach, Florida, for providing line drawings and background material.

Special thanks to my husband, Pentti, for the gift of time; to our daughter Kati, for entering several versions of *Papa Topside* on the computer; and to our sons, Jes, Peter, Eric, and Chris, for their spirited interest while this book was in preparation.

                                                        Helen Siiteri

Introduction

In the 1960s, an elite group of U.S. Navy deep-sea divers tested the startling theory that humans could live on the bottom of the sea. Under the leadership of Navy Capt. George Foote Bond, these first "aquanauts" volunteered for a series of experiments that progressed from a land-based submarine escape training tank to Sealab, a sophisticated habitat deep on the ocean floor. Captain Bond described his aquanauts as "a special breed of men who depend—literally— upon one another for their lives." Flamboyant, arrogant, intensely competitive, these special men brought their unique skills and long diving experience to the two successful Sealab experiments in a display of camaraderie and courage that belied the ever-present risk of failure and inevitable death.

Captain Bond recorded these events during his watch as principal investigator of the undersea living experiments. I edited his personal chronicles and papers to present this profile of the man as reflected by his leading role in the Sealabs. For one exciting decade, the best divers in the U.S. Navy routinely volunteered for assignments under "Papa Topside," a name affectionately bestowed upon Bond and one that delighted him.

This book honors that "special breed"—our aquanauts —and those who watched over them topside.

Helen Siiteri

The Evolution of Diving

900 B.C.   An Assyrian frieze shows men swimming with
               inflated animal skins strapped to their torsos
               for use as air tanks or flotation devices.

               In the *Iliad*, Homer describes the military use
               of divers in the Trojan War.

c. 360 B.C.   Aristotle, Greek philosopher, observes in
               Problemata that ancient sponge divers used
               containers of trapped air to extend diving time.

332 B.C.   Alexander the Great is said to have descended
               in a diving bell to observe his diving warriors
               destroy the underwater defenses of the be-
               sieged island of Tyre.

c. 300 B.C.   Greek laws are passed regulating those who
               dive for sunken treasure.

c. A.D. 77   Pliny the Elder, Roman naturalist, refers to a
               breathing tube in his encyclopedia of natural
               science, *Historia naturalis*. Ancient warriors
               drew air through a tube inserted in their
               mouths while the other end floated on the sur-
               face, an early version of the snorkle.

c. 500   Japanese and Korean women, the ama divers,
               are trained from early adolescence to dive for
               pearls without apparatus or diving dress. (In
               the twentieth century, the ama dive mainly for
               food.)

1240   Roger Bacon alludes to "instruments whereby
               men can walk on sea or river beds without
               danger to themselves."

c. 1511   Leonardo da Vinci sketches an underwater breathing apparatus that encases a diver's head in a leather bag with a breathing tube to the surface.

1662   Robert Boyle, British physicist and chemist, invents the vacuum pump and uses it in the discovery of Boyle's law, which states that the pressure and volume of a gas are inversely proportional to one another. Thus, bubbles of gas dissolved in human body tissues become smaller under pressure at depth and expand as the diver returns to sea level.

1680s   Capt. William Phipps, American adventurer, locates and retrieves 52,000 pounds of gold and silver from a sunken galleon in a salvage operation backed in part by British royalty. Phipps is knighted and later appointed governor of Massachusetts.

Dr. Denis Papin, French physicist and pioneer in the development of the steam engine, suggests pumping a continuous supply of fresh air to a diving bell to extend the duration of a dive. This revolutionary concept evolves into standard equipment for the deep-sea diver: a helmet (hard hat) connected by a hose to an air pump on the surface.

1690   Edmund Halley, the English astronomer, invents one of the most famous diving bells. A weighted barrel is hauled down by ropes to supply fresh air to the diving bell. Halley and four others remain at 60 feet under the River Thames for almost 1½ hours. Twenty-six years later Halley spends more than 4 hours at a depth of 64 feet.

1715   John Lethbridge develops a completely enclosed one-man diving dress: a reinforced, covered barrel of air equipped with a glass porthole for viewing and having two armholes with watertight sleeves. It was intended only for depths shallower than 10 feet.

1776   David Bushnell designs *Turtle,* a small, wooden submarine manually operated by cranks attached to screw-type propellers. During the American Revolution, *Turtle*'s crew planted explosives beneath the British fleet.

1809   Frederic Von Drieberg devises a system in which air is pumped from the surface to a large cannister on a diver's back. Air is delivered to a mouthpiece by the continual nodding of the diver's head.

1819   Augustus Siebe devises an open-dress diving suit in which a metal helmet has an extension in the form of a shoulder plate that attaches to a leather jacket. The helmet, acting as a miniature diving bell, is fitted with an air inlet valve connected to a flexible hose that leads to an air pump. The air is expelled at the bottom of the diver's jacket. Many improvements, including closed suits and telephone connections to the helmets, are made to this dress over the years, but the basic design remains in universal use.

1828   John and Charles Deane market Deane's Patent Diving Dress—a helmet and heavy suit for shallow water. Eight years later they publish the first diving manual.

1864   Benoit Rouquayrol and Auguste Denayrouze design a self-contained breathing apparatus using compressed air in an open circuit (the spent air is forced out). The system stores a small amount of compressed air on a diver's back, so that the diver can disconnect the air hose to the surface and move freely on the ocean floor for a short time. A regulator controls the flow of air to the diver's mouth. This equipment is immortalized by Jules Verne in his classic *Twenty Thousand Leagues Under the Sea.*

1867   Commercial production of air cylinders with regulators begins.

1870s   Paul Bert, French physiologist, discovers the cause of decompression sickness (bends) in tunnel workers and deep-sea divers. Bert attributes this painful and sometimes fatal disease to the sudden change in gas volume when the human body goes from high pressure at depth to sea-level pressure on the surface without giving the body time to eliminate excess gases. Bert advocates recompression (return to the working depth or lower) for victims of the bends.

1878   Henry Fleuss develops a high-pressure bottle of oxygen with a demand regulator. Exhaled oxygen is recirculated through rope soaked in caustic soda, which absorbs the carbon dioxide and purifies the gas to be rebreathed (closed circuit).

1893   The first recompression chamber to simulate rise and fall in air pressure is installed on site to treat laborers stricken with the bends while working on the Hudson River Tube in New York City, at a depth of about 90 feet.

1903   Robert H. Davis assembles the original submarine escape apparatus, consisting of a breathing bag, relief valve, carbon dioxide–absorbent canister, emergency oxygen capsule, main oxygen cylinder and valve, nonreturn valve and flexible tube for charging the breathing bag, and a tube leading to the mouthpiece.

1905   John Scott Haldane, British scientist, expands the understanding of the action of gases under pressure. He devises a stage decompression technique that requires the diver to stop rising every 10 feet for a period of time determined by the depth and length of the dive. The length of time required for each stop, or stage, is listed in predetermined decompression tables for safe ascent.

1915   The U.S. Navy Mark V heavyweight diving outfit features a telephone and improved exhaust valve. Both the Mark V and the British Admiralty six-bolt pattern diving dress are used in commercial projects.

Robert H. Davis develops the concept of using diving bell chambers to rescue crew from disabled submarines.

1917   In the United States, Elihu Thomson proposes that helium, an inert and extremely light gas, be used instead of nitrogen in underwater breathing mixtures.

1920   Joseph Peress builds a diving suit of stainless steel, the Iron Man, which weighs about 800 pounds and functions as a personal submarine. The suit is used in 1935 to help locate the *Lusitania,* sunk off Ireland.

1924   The U.S. Navy's Bureau of Construction and Repair, responsible for Navy diving operations, tries to solve problems caused by compressed air diving by joining with the U.S. Bureau of Mines to conduct experiments with helium as the main gas in artificial breathing mixtures.

1927   U.S. Navy divers descend to 150 feet using a breathing mixture of helium and oxygen.

The first U.S. Navy diving manual is published, with decompression tables to 250 feet.

1928   Robert H. Davis develops a decompression chamber that is lowered to the bottom from a surface support ship. The diver exits to work on the ocean bottom and reenters the chamber through a lower hatch. An attendant assists in removing the diver's helmet. The chamber is raised to the deck of the support ship, and the diver breathes oxygen from special apparatus during the long decompression stages. This technique is used to construct the harbor at Dover, England, and in laying bridge foundations.

1929   Comdr. Charles B. Momsen develops the submarine escape lung used by the U.S. Navy.

U.S. Navy divers go to a depth of 364 feet using a breathing mixture of helium and oxygen (heliox).

1930   Jack Browne develops a lightweight, air-supplied mask for shallow-water diving. The full face mask has an air supply valve and an exhaust valve.

1933   Yves Le Prieur markets a commercial type of self-contained, underwater breathing apparatus (scuba) with a manual control valve that provides up to 10 minutes of air at 40 feet and up to 30 minutes at 20 feet.

1937 Auguste Piccard constructs the first bathyscaphe, a marine application of his free stratospheric balloon with a tightly sealed cabin.

1939 The U.S. submarine *Squalus* sinks at 239 feet off New England. Thirty-three crewmen are rescued in the McCann submarine rescue chamber, a diving bell constructed to lock with the submarine.

Although the first of 640 salvage dives are made with compressed air, most of the dives are made with helium-oxygen mixes. The technique of helium-oxygen diving proves to be far superior to compressed-air breathing for deep diving operations. The Navy establishes 380 feet as the new operational limit for 30 minutes of hard-hat diving on the bottom.

1943 Capt. Jacques Cousteau and Emile Gagnan develop the Aqualung and provide the breakthrough for open-circuit scuba by using a fully automatic compressed-air regulator.

The U.S. Navy's Experimental Diving Unit and School for Deep Sea Divers work out a continuous progression of decompression tables. Although dives to 561 feet are successful, the decompression penalty reaches a ratio of 2 hours' decompression for 1 minute on the bottom.

1959 Comdr. George F. Bond, USN, in conducting research in individual submarine escape, establishes a record for buoyant ascent (no breathing apparatus) in the open sea from a submarine bottomed at 322 feet.

1957–1963 U.S. Navy diving scientists George F. Bond, Robert D. Workman, and Walter F. Mazzone begin the experiments called Genesis in the

Naval Medical Research Laboratory in New London, Connecticut. Bond tests the saturation theory, which proposes that after 24 hours under pressure at a given depth, the tissues of the diver's body will have a gas saturation equal to the surrounding atmosphere. Once saturated, the diver's decompression can be based on depth rather than duration of the dive. Thus, a diver saturated to 300 feet would need the same decompression time (about two and a half days) whether the stay was one day or one month.

1960   Jacques Piccard, son of the inventor Auguste Piccard, and Lt. Don Walsh, USN, dive in the bathyscaphe *Trieste* to 35,800 feet in the Marianas Trench, deepest point known in the ocean.

Hannes Keller, Swiss mathematician, dives to 700 feet in the Navy Experimental Diving Unit chamber breathing a mixture of inert gases. Two years later, Keller descends to 1,000 feet in the open ocean off Catalina Island, California. Although two lives are lost in the dive, Keller accomplishes his goal, proving that humans can dive to incredible depths in the open ocean.

1962   Edwin Link's experiment in saturated diving begins with Belgian diver Robert Stenuit breathing helium-oxygen in a small recompression chamber for 25 hours at 200 feet in the Bay of Villefranche. Using a breathing tube, Stenuit exits the capsule for a short time. His decompression is completed successfully.

Jacques Cousteau launches Conshelf I with Albert Falco and Claude Wesley breathing nitrogen-oxygen for seven days at 35 feet of depth off Marseilles. The divers work a few hours a day at 85 feet, demonstrating that tasks can be accomplished at depth.

1963   The USS *Thresher* sinks with all hands lost in deep water far beyond the reach of the Navy's salvage capability. The Deep Submergence Systems Review Group is appointed to determine the Navy's capability to rescue sailors from disabled submarines and to locate and retrieve large objects lost in the deep ocean.

Jacques Cousteau expands his underwater experiments with Conshelf II and puts seven men in the Red Sea. Five divers live at 32 feet in the main unit, Starfish House, for a month, and two divers live in Deep Cabin at 82 feet for a week, making excursion dives to depths below 330 feet. A prime goal is to test a diver's efficiency when working extended hours from a pressurized habitat. The French oceanauts also demonstrate that a saturated diver can descend to a lower depth for a reasonable period and return without difficulty.

1964   Edwin Link and Dr. Christian Lambertsen send Robert Stenuit and Jon Lindbergh down for two days at 432 feet off the Bahamas. After the divers complete their assigned tasks, they surface in their submersible decompression chamber, which couples to a large recompression chamber on the deck of the support ship. The divers transfer to the deck unit, where they decompress safely for nearly four days.

The U.S. Navy approves a five-year program to achieve deep ocean capability. Capt. George Bond and his coinvestigator, Capt. Walter F.

Mazzone, coordinate Sealab I and send four Navy divers down for eleven days at around 193 feet off Bermuda. The aquanauts breathe a heliox mixture and demonstrate that no short-term physiological damage occurs from the extended dive and short decompression.

A hooded life jacket developed by Lt. H. E. Steinke, and tested by Steinke and Comdr. Walter Mazzone in open-sea ascents, is adopted as standard apparatus for submarine escape.

1965   Prior to June 1965, efficient decompression tables were not available for helium mixtures in scuba diving. Under the supervision of Capt. Robert D. Workman, MC, USN, tables are developed over a period of years and tested in open-sea trials. Prior to the trials, sixty indoctrination dives are made by the Navy's Atlantic and Pacific diving fleets. The introduction of new diving tables is completely successful.

Capt. George Bond, principal investigator, and Capt. Walter Mazzone divide twenty-eight divers into three teams that spend fifteen to thirty days at 205 feet in Sealab II off La Jolla, California. Astronaut–aquanaut Scott Carpenter serves as team leader for two shifts, and Master Diver Robert C. Sheats leads the third shift. Extensive physiological data are collected from the aquanauts, and complex job assignments are performed on the bottom.

Simultaneously, in Conshelf III, Jacques Cousteau puts six divers for twenty-two days at 328 feet off Monaco in the Mediterranean. The goal is to prove that divers can perform difficult maintenance tasks over a period of time at that depth.

1970s   Tektite I (1969) is sponsored by the U.S. Navy, Department of the Interior, and National Aeronautics and Space Agency (NASA). Four divers submerge for sixty days in 40 feet of water off St. John, U.S. Virgin Islands. Extensive biological studies produce a wealth of scientific data, and behavioral studies prepare for future journeys into space. After sixty days' submergence, the aquanauts decompress for 19 hours on a support barge.

In Tektite II, eleven successive five-person teams spend fourteen to twenty days at 50 feet off St. John. One all-female team of scientists, led by Dr. Sylvia A. Earle, prepares the way for women in space.

In AEGIR, sponsored by the National Oceanic and Atmospheric Administration (NOAA), six aquanauts stay for six days at 520 feet off Hawaii.

Graham S. Hawkes, engineer, designs the JIM suit, a one-man atmosphere diving system (OMADS). The JIM suit is named after Jim Jarratt, the original wearer of the Iron Man armored suit. It is lowered on a tethered line from a ship or platform using a winch and cable.

1979   Dr. Sylvia A. Earle, oceanographer, walks for 2½ hours at 1,250 feet on the ocean floor off Oahu, Hawaii, wearing a JIM suit, establishing a record for the world's deepest untethered dive.

1980s   Graham S. Hawkes pilots the one-person submersible *Deep Rover* to 3,000 feet, a world's solo dive record.

In the U.S. Navy, researchers achieve voice recognition in diver-to-surface communications using helium unscramblers from 1,800 feet in depth.

Solid-state video cameras, hand-held or mounted on the diver's helmet, provide broadcast-quality color images of the underwater site to topside viewers who monitor the diver.

Arctic climate diving is made tolerable by wearing suits designed to circulate heated water over the body through capillary lines, while a small heat exchanger warms the gas the diver is breathing.

A need reiterated by Capt. George Bond during the Sealab experiments is met with development of a device to track a free-swimming diver and to precisely locate objects on the ocean floor. The diver wears or positions an electronic pinger-receiver, which ties into a long-range navigation (LORAN) system. A flashing light indicates the latitude and longitude of the diver/object on a grid pattern.

In the U.S. Navy Experimental Diving Unit's hyperbaric chamber at Panama City Beach, Florida, a computer monitors the percentages of oxygen inhaled and carbon dioxide exhaled by a diver in a breath-by-breath analysis. The diver's core temperature is also tracked.

This chronology is based in part on an outline developed by The Museum of Man in the Sea, 17314 Back Beach Road, Panama City Beach, Florida. The sequence is illustrated by Paul Grandinetti in a series of original watercolors on permanent display in the museum through the courtesy of Battelle Laboratories, Columbus, Ohio.

Papa Topside

Prologue

After six long years of scientific research in my lab at the U.S. Naval Submarine Base in New Haven, Connecticut, I was still regarded by many as a rank newcomer to deep-sea diving medicine. In retrospect, our first underwater experiments with deep-sea divers might never have progressed to the Sealab projects if it had not been for sudden recognition of the Navy's need to extend its diving depth. But when the USS *Thresher*, the Navy's lead nuclear attack submarine, and her crew of over one hundred men and seventeen civilian observers were lost on 9 April 1963, outraged Americans demanded that the Navy provide a way to rescue the courageous men serving in our nuclear submarine fleet.

I was in my office at the sub base that tragic day in April when my phone rang. It was Comdr. Bill Rothamel, executive officer, with a terse message that sent a chill pulsing through me.

"George, *Thresher* is reported missing and may be sunk. Hold your best divers aboard. We may get a chance to rescue those men."

I assembled our diving tank instructors and told them to stand by for an emergency. We held desperately to the hope that *Thresher* had landed on the continental shelf, giving us a marginal chance for rescue operations. But we soon learned that the sub had sunk far below the crush depth of her hull and the survival limit of her crew. All persons aboard, many of them old friends and shipmates, were lost. My wife, Marjorie, who shared these friendships and cherished memories with me, immediately called to offer whatever comfort she could. There was nothing anyone could do at that point except grieve for the dead and honor them.

Official reaction in the top echelon was immediate. As the Board of Inquiry convened, Secretary of the Navy Fred Korth ordered Rear Adm. E. C. Stephan to assemble a group of undersea experts to form a committee with almost limitless powers. Its mission was to call on the best minds and physical resources of our nation to establish directives for development of undersea and surface support systems to prevent, or at least recover from, any future disaster comparable to that which befell the *Thresher*. Edwin A. Link, the American aviation electronics pioneer and inventor, was a key member of the committee, for as Admiral Stephan noted, "If you get Ed Link, then the others will fall in behind."

After nearly eight months of review, this committee, entitled the Deep Submergence Systems Review Group (DSSRG), recommended that a five-year program be initiated in the Navy to tackle deep-sea salvage of large objects, recovery of submarine personnel, and extension of diving operations to include useful work from an undersea habitat. The last aspect would be called the Man-in-the-Sea program.

Parts began to fall into place when the Office of Naval Research (ONR) became interested in my proposal to conduct sea trials of the results of our laboratory research on the human capability to live and work on the ocean floor. Our project would fit into the new Man-in-the-Sea program and would also fill an acute need for exploitation of the basic physiological studies that made long-duration deep diving possible—almost all of which had been done under U.S. Navy sponsorship. It was somewhat ironic that the Navy had made no effort to apply this research to the ocean and, indeed, seemed to have been left at the post by the energetic, well-publicized efforts of my friends Jacques Cousteau and Ed Link. I must note that while sharp competition existed amongst our three separate groups, we freely exchanged information and scientific data.

Who was I to share scientific data with the legendary Cousteau? We first met in New York, backstage at Town Hall at an international seminar on undersea activities when I was just starting our medical research project in diving. We talked our way from the speakers' platform to the apartment of Cousteau's good friend, writer James Dugan and his wife Ruth, where we all sat and talked until sunrise. Cousteau was looking for a safe and practical way to keep divers submerged for an extended period without paying the high penalty of time spent in decompression. I told Jacques about my first experiments to test the saturation diving theory and we agreed to stay in touch.

It was obvious to all of us engaged in the development of deep-diving techniques that it would be to a diver's great advantage to remain on the bottom long enough to complete any work at depth. I proposed that decompression time increases with depth and duration of the dive up to the point where the diver's blood and tissues have absorbed all the gas they can hold at that depth. At this saturation point, the time required for decompression remains fixed, and the diver could work at that depth, if given a shelter to retreat to, for months instead of minutes.

At that time, conventional Navy diving practices called for the diver to descend from the surface to a working depth of perhaps 190 feet for a period of 30 minutes, and then return to the surface to spend more than an hour in a decompression chamber. This procedure was satisfactory for brief, shallow bottom stays with short periods of work but was totally inadequate for exploration of the continental shelves, which slope to about 600 feet.

We proposed that the undersea worker be provided with a sea-floor house pressurized to ambient pressure and filled with a gas mixture suitable for breathing. From this habitat, the diver could exit directly into the ocean for many hours of useful work and then return to the safety and comfort of the underwater house. Since the diver would experience no significant change in pressure between the habitat and the sea floor, there would be no decompression penalty for these excursions. Instead, the decompression of the saturated diver could be accomplished all at one time when the work was finished and the diver returned to the surface. This concept, if it worked, seemed to offer the needed breakthrough in manned undersea venture.

After years of struggle to obtain the necessary support, my coinvestigator, Comdr. Walter Mazzone, and I had shown that there was no evidence of health damage in animals exposed to a depth of 200 feet for twelve days while breathing a helium-oxygen mixture. We were then allowed to conduct physiological and psychological tests in simulated conditions on U.S. Navy divers. Our tests showed that after 24 hours at any depth of water, a diver becomes wholly saturated with the gases that make decompression necessary and the decompression penalty can be paid in one single linear ascent. We dubbed this project Genesis and determined to conduct our work in an orderly, carefully documented, scientific manner that would progress to our goal of placing men and women on the ocean floor to live and work freely.

Ed Link's expertise attracted funding from various sources that made him one of the first to launch a human in an underwater living test in the open sea. In September 1962, under Link's watchful eye, the Belgian diver Robert Stenuit reclined in a small recompression chamber and was lowered to a depth of 200 feet in the Bay of Villefranche, France. Stenuit breathed a helium mixture and exited for several dives, using a breathing tube, before his 25-hour experiment was over.

Meanwhile, Jacques Cousteau was conducting his first underwater tests in saturation diving by placing his long-time colleagues Albert Falco and Claude Wesley at a depth of 35 feet in the Mediterranean off the coast of Marseilles. They lived for one one week in an underwater house, exiting through a hatch in the bottom and descending to 85 feet to perform simple tasks. Compressed air was used in this test and although the men experienced some discomfort, they reported no ill effects. Less than a year later, for Conshelf II, Cousteau planned an underwater colony off the Sudanese coast in the Red Sea. Five oceanauts lived for a month at 32 feet in depth, and two men went even deeper for a week. Cousteau wrote in *The Ocean World* (New York, 1979), "Soon the men living in the Conshelf station realized there were subtle effects of the pressure. Beard growth was retarded. Small cuts and abrasions healed much faster than similar injuries suffered by our divers living on the surface." Toward the end of the month, one French oceanaut noted in his diary that he was just beginning to be aware of the passing of time. "I feel I may rise to the surface next week, without having seen and experienced absolutely everything," he lamented. This urgent concern to explore the vast richness of the undersea world was to be voiced by many oceanauts and aquanauts in the months to come.

Link then fabricated SPID (submersible, portable, inflatable dwelling) and submerged Robert Stenuit and Jon Lindbergh in it for two days at 432 feet in the Bahamas. On the last day of June 1964, the two young men descended in the submersible decompression chamber that served as an underwater elevator to the SPID and 49 hours later emerged, as Link put it, "home safe." This was a remarkable advance, as no one could be positive of the effect of sustained pressure of 200 pounds on every square inch of the human body.

Three weeks later, our four Sealab aquanauts, Bob Thompson, Tiger Manning, Bob Barth, and Andy Anderson, stayed for eleven days at around 193 feet off Bermuda. The aquanauts achieved all our experimental objectives, did some useful work, and dramatically advanced the potential of human beings to live and work on the ocean bottom.

In 1965, we progressed to Sealab II, where twenty-eight of our aquanauts, divided into three teams, remained for fifteen to thirty days at 205 feet. The teams accomplished a wide range of salvage and undersea construction tasks at depth, with no significant short-term physiological changes recorded. Simultaneously, in Cousteau's Conshelf III, six oceanauts, living at 328 feet off Monaco in the Mediterranean, ruefully concluded that "machines fail more often than men."

These events opened up a chapter of diving advancements during a very short span of time. The need to advance from the old plateau of around 200 feet in diving capability might be reckoned from the launching of our first nuclear-powered attack submarine in 1955. The USS *Nautilus*, under Comdr. Eugene P. Wilkinson, operated routinely at a depth far below our rescue capability. This challenge was not fully met until after 1963, when the *Thresher*'s fate compelled the Navy to provide technological support for deep-sea search, rescue, and salvage. Dr. Robert Workman, Navy physiologist working with the Experimental Diving Unit, Walt Mazzone (my coinvestigator), and I suddenly found our Sealab project on the priority list. A new age of exploration had arrived and we stood on the frontier.

1

A Submariner
(1953–1957)

When Walt and I scheduled our visit to the team of aquanauts living in the Sealab II habitat early in October 1965,
we knew it would be our last dive to the Sealab II site. It
might well be our last dive together to any Sealab site, since
the next stage of the project would be going deeper, and we
were growing older. I felt that this trip had a special significance—not as a farewell to diving, but as a last look at a
scene I had come to hate and cherish with equal intensity.
That day I would learn to fear it as well.

Walt and I agreed that when we reached the sea floor we
would not enter Sealab II, moored on the bottom, but
would just stick our heads up into the entrance hatch, shoot
the breeze with our aquanauts, then duck down for a leisurely inspection of the Sealab II hull, take a few photos,
and make a slow return to the world of sun.

Early in the afternoon, Walt and I were helped into our
wet suits and diving gear: double tanks of 75–25 helium-
oxygen mix, weights, knives, compasses, watches, cameras,
and a buddy line. We flapped awkwardly to the diving station and rolled off to swim to our diving bell, scheduled to
take us to within 4 fathoms of the Sealab II habitat. Inside
the bell, we checked our gear and passed the word to lower
away.

With the whine of the lowering cable, Walt and I dropped
quickly through the upper, lucid layers of the Pacific. We
saw schools of anchovies, and other unnamed fish hordes,
which approached and even appeared to attack our open

bell, only to disappear from sight as we lowered into the black, colder void below. In less than 2 minutes, we had reached 180 feet, and the free ride was over. A long scan of the underwater scene revealed a dim light below, at an azimuth unknown to either of us. It had to be Sealab II, but which end or which side we did not know. Walt nodded to me and I responded with a thumb down. We started our free dive downward, independent of the descending line, which was fouled on an object far above us.

Down, straight down we swam, suddenly aware of the syncopation of our fin-beats and exhalation sounds, so loud in the silent, black world. About 2 fathoms off the ocean floor, we suddenly broke free of the murky cloud through which we had been descending. Below us, the bottom was free of all turbidity. Visibility jumped from inches to many feet, and Sealab stood clear in its entirety. Light seemed to emanate from the ocean bottom, while above us all was black. On a mutual impulse, Walt and I rolled on our backs to stare up at the world from which we had come. It did not exist; only a forbidding black curtain lay above. Only below was there light and safety. Surely this was our refuge, and our home. Suddenly we stared at one another, each reading the identical thoughts of the other, disoriented in space, time, and philosophy. This, we recognized for the first time, must be the "breakaway phenomenon" of undersea existence, a profound revelation for the divers, a frightening challenge to the topside watch-standers.

The spell was abruptly broken by a sharp pressure on my right arm. I spun about to face one of the aquanauts, sent out to save us from wandering past the security of the Sealab vicinity and the limited range of our compressed breathing mixture. We swam slowly to the shark cage, guided by our guardian aquanauts, to whom this part of the continental shelf was a familiar front yard. Graciously sweeping aside the poisonous scorpion fish from the steps, these undersea experts and friends brought us to the only safe area of 2 square feet for miles around—the entrance hatch to Sealab II.

Thrusting our heads through the Alice's Looking Glass of the undersea habitat, we were greeted by a ring of friendly faces at the periphery of the hatch. At the moment, I could not help but reflect that all aquanauts have countenances, rather than faces. We exchanged banalities, the undersea veterans suppressing laughter at our squeaky voices, which sounded like Donald Duck in the helium atmosphere to which we were unadapted. A word or two, a handshake all around, and we were ready, with our limited gas supply, to return to the refuge of our diving bell—followed by nearly 90 minutes of decompression to pay for our brief visit to such a depth.

According to our prearranged plan, Walt and I were to part company briefly for individual photography, then meet in our diving bell. I was to swim along one side of Sealab II, while he covered the other. After one sweep, approaching the shark cage, I spotted a nylon line that angled toward the blackness above. I concluded that this line had been rigged by an aquanaut as a guide to the diving bell. Blindly, I began an ascent along my lifeline, confident that a few fathoms above I would find refuge and a breathing supply to replace the dwindling gas in my bottle.

I found neither. As I ascended the line, no diving bell came into sight; for that matter, nothing was in sight, since I had again penetrated the black cloud. Venting of my middle ears and common sense told me that I had ascended several atmospheres—far above the level of the bell. Now my breathing gas was nearly gone, and might not last if I tried to retrace my route. However, I started back down the line, convinced that my gas and, finally, my luck had simultaneously run out.

As I broke through the black strata of water, an unseen

hand seized my flipper and another took my arm. A lateral tow of 10 yards, an upward thrust, and I swam the last fathom to the diving bell and safety. Walt was waiting and anxious. We started our ascent immediately as I tried to explain my inexcusable blunder. Then I lapsed into silence while we completed the oxygen phase of our decompression. Topside, Walt and I undressed, showered, and ate in introspective silence, then took our watch stations. About an hour later, Walt came over the intercom: "That was one weird run and you're one lucky diver, George!"

It was a strange and wonderful day. Only now does it occur to me that I made no gesture of thanks to the aquanaut who saved my life, and I never found out who it was that guided me to safety. In my heart, I know that the whole damned team did!

There was little in my medical training or experience prior to my career in the Navy to prepare me for underwater exploits. My practice of medicine was centered in Bat Cave, North Carolina, a small community midway down Hickory Nut Gorge in the Blue Ridge Mountains. In 1946, Bat Cave numbered about 170 souls—give or take a dozen; there were thousands more tucked about in the quiet pockets of the Blue Ridge. My goal was to establish a rural medical practice and, eventually, to build a hospital.

As a boy, I had explored the fascinating rock faults found in the Bat Cave area. The interior of the cave is evenly tempered in all seasons by an underground river that we could hear but never reach. For a few months each year, the cavern quarters a horde of bats that migrate from the Black Forest of Germany. Their presence was not generally appreciated by the local community. They knew that a cave full of bats was up there on the side of Blue Rock Mountain, but few actually climbed up to verify the fact.

As a youngster, I was welcomed into the remote cabins

around Bat Cave through our family's close ties with Uncle Beaufort Heydock, who tended the animals on our newly purchased Chestnut Gap farm. I often listened to the beguiling legends and lilting folk songs preserved down the generations in the Scots-English lineage of many of the North Carolina mountain settlers. I knew the surrounding hill country. The gorge and Rocky Broad River had been my favorite camping area for many years. As an overgrown adolescent, I had spent solitary summer weekends camping in Posey's Cave, far up Broad River. I used local personalities as a resource for my master's degree thesis on speech patterns of the people of Appalachia. Yet when I returned in 1946, I sensed that I was a stranger to the people. The difference between being a student philologist and a practicing physician was considerable.

I had entered a premedical program at the University of Florida at Gainesville, then stayed on to earn a graduate degree in philology, with the possibility in mind of obtaining a Ph.D. so I could teach my way through medical school. Although I missed this goal by a short margin, I did get my master's degree in language and English literature. I was lucky enough along the way to marry a wonderful girl, Marjorie Barrineau, and we left, finally, for McGill School of Medicine in Montreal, Canada, where I was accepted in 1941.

Immediately after Pearl Harbor, I applied for a reserve commission in the Medical Administration Corps, U.S. Army. This commission was awarded in March 1942. As a medical student, I was then assigned to Ready Reserve, subject to active-duty call on fifteen days' notice, and allowed to continue my medical education.

In my first year of medical school, I found I had been too long away from the basic sciences and had to repeat much of this work to rise to an acceptable level of competence. The next year, following the birth of our daughter Abigail, I entered Montreal's Homeopathic Hospital as a student intern and began to learn something about medicine and surgery. Although I had to cut classes frequently to meet my ongoing commitments at the hospital and to teach a course in English literature, my grades gradually improved. The daily exposure to clinical medicine at Homeopathic made up for the lapses in classroom instruction. I learned that every good physician is a teacher at heart and will take time to pass along improvised techniques to an aspiring general practitioner.

In May of 1945, I completed my rigorous final examination at McGill with creditable grades and went on to North Carolina, where I started a rotating internship at Charlotte Memorial Hospital. Meanwhile, a son named George became a welcome addition to our family. As a freshman intern, one of the five available to assist with treatment of more than five hundred patients, I learned at once the routine of a regular 18-hour day, which frequently grew to 20 hours or more. My days and nights at Homeopathic stood in good stead as chronic fatigue became a way of life. Charlotte Memorial was a relatively new five hundred–bed hospital, overloaded with patients and critically undermanned by professional staff. To compound the strain, we had inherited from nearby counties about 150 youngsters severely ill with acute poliomyelitis. I became so caught up in my duties that pending military obligations were forgotten.

The Pentagon did not suffer such a memory lapse, however. In February 1946, I received an official letter acknowledging that I was now a medical school graduate and, therefore, a candidate for a commission as first lieutenant, Medical Corps, to replace my present status as second lieutenant, Medical Administrative Corps. I was further ordered to report to Fort Jackson, South Carolina, for physical examination. In deference to the authoritarian tone of the letter, I reported to that Army post.

Over the years, I had sustained a skull fracture, a herniated disc in my back, and a collapsed left lung. Although these events had occurred during my tenure as a medical student and Army Reserve officer, I had not reported them to higher military authority. This oversight was quickly uncovered at Fort Jackson, when my medical history was revealed. In May came a shocking letter from the Secretary of the Army that acknowledged my new professional status and tendered a firm commission in the Medical Corps, subject to my signature on a document of waiver of physical disabilities.

With utter disbelief, I read the conditions of this waiver: "This officer, upon acceptance of the enclosed commission, agrees to waive any and all conditions of the head, chest, or back, for the purposes of subsequent claims of medical disability." This struck me as unfair. Under these conditions, I could only claim a line-of-duty medical disability if I shot myself in the foot or otherwise became maimed below the navel. I returned the commission and waiver unsigned, declining the offer with a negative commentary. Somehow, I could not feel unpatriotic, since VJ Day had just been celebrated and the war was over. Three weeks later, I received an official letter to the effect that, in the light of my negative response, the commission had been withdrawn. With a sigh of relief, I continued my hospital chores through May 1946 and made plans to start a country practice centered in Bat Cave.

From my first day of practice, I devoted a disproportionate amount of time and energy to house calls—though I preferred the term "family practice in the home." To offer my best under these circumstances, I carried with me, packed securely in my war-surplus Jeep, a complete set of sterile surgical and obstetrical drapes and instruments, plus a load of drugs for dispensing to patients. In this fashion, I sought as best I could to bring modern medical care into an eighteenth-century home environment.

All the while, I knew full well that I was only marking

time, although I dispensed a reasonably good grade of medicine at the meager facility I dared called the Valley Clinic. I used the Hendersonville hospital, fifteen miles distant, for all my major surgery and many first childbirths. But these facilities were stopgaps at best. Home visits—which called for almost 200 miles of day-and-night driving over nonexistent roads, long hours in wait for difficult home deliveries, and emergency surgical procedures performed under unbelievably primitive conditions—were a large part of my practice. As the months passed, I fell into a routine of 18-hour days, traveling too many leagues and practicing essentially three separate grades of medicine: home visits, obstetrics, and surgery. But if at first this pattern was distasteful, it was educational beyond belief. Much as I was distressed by the primitive conditions I found in the shacks, cabins, and houses in my 400-square-mile territory, I developed an indelible picture of my patient-families. In that period of time, I wore out two Jeeps, aged more than a little, and shamelessly neglected my burgeoning family, which by then included a second daughter, Judy, and eventually the last child, David.

One night in mid-January 1947, about a week after an influenza epidemic hit nearby cities, I saw my first case, in a cabin on Rich Mountain. It was an unusually severe occurrence of Spanish influenza, in the midst of a large family. Somehow, I missed the implication of this viral hazard. Relying on the basic isolation of mountain families, I thought—naively—that the virus would not be dispersed. How wrong I was! Mountain families tend to help one another. In the course of neighborly assistance, germs are passed that can grow into an epidemic, which is exactly what happened. The moment of truth dispelled my complacency. I had grown to love the pattern of the practice because I loved the people and their way of life, and felt I could handle their health needs. The influenza epidemic returned me to reality.

As soon as I recognized the certainty and scope of the epidemic, I closed the Valley Clinic and turned my Hendersonville hospital practice over to my city colleagues. I was committed to a program of total home care. How long it would last, I could not guess. I learned to do a great deal with very little in the way of resources. As my supply of fever-reducing capsules diminished, I searched the cupboards for Stanback powders or other home remedies for fever. For the devastating coughs that developed with the onslaught of the sickness, I ferreted out home supplies of corn liquor and the juice of canned peaches. A combination of the two, watered down and with the therapeutic addition of codeine, worked quite well.

Even in moments of fatigue, chill, and despair, I felt a strong reliance on the power of prayer. My patients invariably sensed this and almost always asked me to intercede in this way. Such an ingredient is important in the healing process, and I did not hesitate to use it liberally.

Before the epidemic ended, over nine hundred persons were stricken. But I was only one physician dedicated to the medical care of about six thousand people, widely scattered over a large and rugged area. When the epidemic was finally over, I realized that my routine of constant travel was clearly not a solution to the health needs of my patients and, if I continued, my life span might well be no longer than that of my Jeep.

A few more weeks passed before I realized I did not feel well and went home with a temperature of 103° F, figuring I had influenza. As the days wore on, I began to experience symptoms that were not compatible with a diagnosis of influenza. A constant headache was a puzzle, as I had never experienced one before. There was a pain high up on my left side. Marjorie called our friend Nick Fortescue and asked him to make a house call.

After exploring the upper parts of my anatomy, Nick

turned his attention to my abdomen. As he worked over the left side of my belly, he grunted with amazement: "God-damn it, George, you've got a spleen the size of a basketball! You'll go to the hospital tomorrow."

Next morning in Hendersonville, I submitted to the inevitable bloodletting and collection of body wastes. One at a time, my colleagues came by to examine me. Obviously, not one of them liked what they found, but they withheld diagnostic judgment until the lab reports were in. Since the lab results were a focal point of diagnosis, I asked for an interview with Lorraine Brooks, the X-ray technician. She stopped by my room in the early evening, bringing her films and the blood studies done earlier in the day. The radiographs clearly showed a massive enlargement of the spleen and of the glands midway in the chest. That was not news to me, but the hematology report was a shock. My hemoglobin, normally about 15 grams, was down to 11. My white cell count was approaching 100,000, against a norm of 5,000. A differential count showed most of the cells to be myelocytes. The picture was crystal clear. I had myelogenous leukemia and could look forward to a life span of about three months.

The next morning I was the subject of Medical Grand Rounds. A dozen close friends on the staff gathered in my room. Nick Fortescue began the dialogue: "Goddamn it, George, you look about as ready to talk as you'll ever be. Want to hear us out?"

"I think not," I replied. "The nausea, the glands, the spleen, and now the blood. I don't need to spell it out any further. Acute myelogenous leukemia is a damn sight easier to diagnose than mumps and has six fewer complications."

Nick cleared his throat. "The final diagnosis is complete only when a pathologist looks at your bone marrow and says, 'That's it!' Then, and not before then, do we accept your diagnosis. You know that as well as we do, George." My colleagues were asking me to go to one of the university hospitals at the other end of the state for ultimate diagnosis and perhaps a trial treatment.

"I can't spare this time away from my family," I said. "If we're right in this diagnosis, then a pathologist can confirm our estimates at autopsy. If we're wrong, I'll survive. In either case, I need time with my family. Please release me today to go home." The next day I was transported from the hospital to my bed in the log cabin on the hill. I did not expect to recover.

For the next six weeks I stayed in bed, exhausted and steadily losing weight until I was down to 160 pounds, against my usual weight of over 200, and well below the norm for my 6-foot 3-inch frame. The six-week bed rest gave me many good hours shared with Marjorie. Removed from the everlasting emergencies of medical practice, I was able, at last, to admit to myself that since mountain people had survived their hardships before my arrival, they would continue to do so regardless of my personal professional attendance. The realization that I was not their messiah was exhilarating.

Before the end of those six weeks, Marjorie and I also realized that we had become hopelessly insular in our thoughts and interests. We began to examine matters outside our small, intimate world with new enthusiasm. And as the weeks went by, I grew less and less concerned about our future, there or in a different location. It was enough that our family was together and loved one another very much.

When a radiologist friend telephoned to pass along information on one of my patients, he apologized for his tardy report, saying he had been on the sick list for some weeks. Almost offhandedly, he remarked that he had seen my X-rays and that we shared the same medical problem, although I must be somewhat recovered by now.

"I'll be on half-time for a few months, but infectious mononucleosis is a self-limited disease, after all," he said.

When he asked what my heterophile ratio was, I was too stunned to reply but muttered my thanks and hung up.

Early the next morning I was in the laboratory to take the one blood test I had overlooked: the heterophile agglutination test for the nonfatal illness commonly known as the "kissing sickness." I learned from the test that I was not dying of leukemia; I was alive with infectious mononucleosis and, in fact, I was suffering mostly from a swollen spleen and incomplete self-diagnosis. As soon as I was able to get back to work, I began the first move to get a hospital built for my mountain patients.

In April 1947, I called a meeting of about a dozen men, each a leader of a community within my broad area of medical care. We met in the vast living room of the Esmeralda Inn, a venerated structure dating back nearly one hundred years. Attendance was 100 percent. I made it clear that my efforts to practice medicine under archaic conditions could only result in my death or desertion in a matter of years, with no replacement in sight. Given an institution in which modern practice could be assured, we could do better by our people, and ultimately attract other practitioners, who would take my place to guarantee continuity of care for the foreseeable future. In all honesty, I had to point out the problems of expense, impossibility of state or federal government assistance, and lack of trained personnel, as well as the need for total commitment.

One at a time, I polled the dozen men who would speak for nearly six thousand souls. One by one, these remarkable people gave a verdict that was unanimous. We would build a hospital—the Valley Clinic and Hospital, a nonprofit, community-owned institution. The mechanics of its construction and administration would be clarified at a meeting one month hence. We shook hands and each returned to his bailiwick, to talk to people. For my part, hours of solid planning lay ahead, but we were off to a good start and I was confident there would be no turning back. From that point to this day, the future of the Valley Clinic and Hospital has never been in doubt. If collusion was involved at any turn of the many difficult steps on the way, it was for the best of possible causes, and could never be proved in any event.

On a bright day in mid-October 1948, the Valley Clinic and Hospital became a reality. Standing on a podium built by my best friends, Fate Heydock and Lonnie Hill, I presented the hospital deed to the board of trustees. In the local bank, our cash reserves amounted to less than $1,000, but the hospital was open and prepared to receive patients. Inside, I already had a patient, a member of the Pitillo family in labor.

The imminent birth was especially heartwarming since, in the beginning, I had been persuaded to become a physician at Bat Cave by Duga Conner, a superlative midwife and friend. A tireless, dedicated woman, she fully answered my questions as an ignorant youngster about the nature of her duties. In a matter of minutes, she clarified her role as chief midwife of the area and told me, in understandable words, the vital details of human birth. Then this wonderful woman turned to me and said, "George, why don't you get the schooling to be a doctor and then come back here? We've never had nary a one and we need one bad." I promised I would do as she asked. Hers is the example on which I base my esteem of obstetrical practitioners. At times, this practice calls for the diagnostic experience of a good clinician in combination with the skills of a surgeon. I have learned that childbirth is a hazardous process, with surprises from conception to delivery, and when a doctor can in any fashion assist in the birth process, he best fulfills the role of "beloved physician."

During my seven years as family physician on the mountain, I often saw the other side of life's coin. In the final days of my country practice, I was called to Remus Lytle's house, high on the upper shoulder of Bald Mountain. It was far past midnight when I brought my Jeep to a stop in the front yard of the cabin, but the fatigue I had felt earlier had long since been washed away in the clear mountain air. No matter the time of day, I was about my chosen business. A single kerosene lamp inside the cabin cast soft shadows of the huge boxwoods beside the porch, and to the west I had a rare view of the full moon setting. There was no sense of urgency as I gathered my small bag of medical instruments. I was not on my way to treat an illness but to bid farewell to an old friend and patient, Aunt Silla Lytle.

Remus and his wife, Pearl, met me on the porch with quiet words of thanks and ushered me into the living room, where my ninety-year-old patient lay on the sofa, properly outfitted in her Sunday dress, with a light coverlet pulled up to her chin. Her eyes were closed but she was not asleep, and when Remus whispered, "He's here, Aunt Silla," the lids opened wide as she sought my face through the curtain of cataracts that blurred her vision. As I sat beside her and took her hand, she spoke in a clear whisper.

"It was good of you to come, doctor. This time, you've no call to use your heart-listener and pressure machine." She paused a bit, searching my face with those nearly sightless blue eyes, then delivered her message.

"You've doctored me goin' on seven years and fer that I give ye thanks until you're better paid. I've said my prayers, and yours will be to say at the burying tomorrow." Then she paused, to gain her last ounce of strength. At last, she gave my hand a gentle squeeze and murmured almost inaudibly, "Doctor, it's gettin' bright light and black dark all at the same time." With that mystical announcement, Aunt Silla Lytle expired in an aura of peace that does not often come with death. My useless stethoscope remained in the leather bag, not needed for her silent heart.

At last Remus broke the silence: "Least I kin say, Doc, is

that Aunt Silla never died of no disease. She just plain punied out."

As I helped the couple prepare the body, I pondered those words and wondered if one day I would have the professional courage to write on a death certificate the simple and adequate phrase "Just plain punied out."

For no accountable reason, I cranked up the Jeep and drove up the rocky cowpath to the peak of Bald Mountain, a mile beyond the cabin. At the bare summit of the peak, I cut the engine, lit my pipe, and paused awhile to do some thinking. A half-mile below, the dim outlines of Hickory Nut Gorge were faintly visible, lighted only by the setting moon to the west and the coming easterly dawn. Towering over the gorge were Chimney Rock and Sugarloaf Mountain, with Bearwallow and Little Pisgah guarding its high reaches. Thin shrouds of fog covered the lake below and, far to the east, the early rays touched the monolithic spire of Table-rock Mountain.

In this majestic setting, I sat in the Jeep and surveyed the mountain land to which I had assigned myself. Scattered at random amongst those hills and rock cliffs were some six thousand people, all friends, and all patients in times of illness. Out of sight, in the dark shadows of the gorge, lay our Valley Clinic and Hospital, and downstream on the Rocky Broad River rose the Church of the Transfiguration, both built by mountain people.

The sun began to edge up over the low ridges of Polk County to the east. This was the Sabbath and my day to conduct services in our church. I eased the Jeep into low gear and drove past the Lytle cabin to the church far down the mountainside and then to the rectory residence of Father Frank Saylor, the priest in charge of our Church of the Transfiguration, and my close friend of many years. Single-handedly, Frank Saylor had salvaged vital relics from our early church as it burned to the ground, and he was now completing construction of a new structure that would in time become the most beautiful church in my experience. On many occasions, we had engaged in long conversations concerning some of the unclear areas of theology. Father Frank was a rigid adherent to the smallest rubrics of our Episcopal church ritual, but welcomed my assistance in delivering the homily at Sunday service. My life was filled with the satisfaction of going about my chosen work in a place I wanted to be, supported and surrounded by loving family and loyal friends.

However, with the onset of hostilities in Korea in June 1950, my time at Bat Cave ran visibly short. On the day following our announced military support in that struggle, I received a brief telegram from the Third Army alerting me to my active duty status and eligibility for recall for overseas assignment. This development was somewhat unexpected, as I had considered myself beyond the pale of military service, but I was recalled in due time.

I agreed in 1953 to relieve an Army paratroop doctor, providing that that doctor would relieve me in Bat Cave. The turnabout took place in fewer than 14 hours, and I was on my way. At daybreak on 3 September 1953, I drove my Jeep through Bat Cave, past the hospital, and over the top of Hickory Nut Gap, headed for Fort Sam Houston in San Antonio, Texas. I did not look back.

The conflict in Korea had ended at Panmunjom in July 1953. In September, after less than a week of close-order drill at Fort Sam Houston, I joined nearly three hundred of my fellow recallees to hear that the Army had made an oversight; we would be absorbed instead into either the Air Force or Navy, where doctors were needed. I opted for the submarine service, since it paid the identical hazard compensation offered by the paratroop service for which I had originally contracted. Five days later, I found myself at the U.S. Naval School of Deep Sea Divers, not exactly on my planned agenda. I was not only surprised to be there but also delighted to find that I loved to dive. This fascination with diving has stayed with me all my life. After training at the diving school, I moved on to submarine school, where I learned to love submarines as well.

However, any officer of our armed forces inevitably encounters the oddball who loses sight of the ultimate military goal and shows a disturbing tendency to pick daisies during the forced march. This man is not a candidate for the higher echelons, but he derives an immense store of pleasure during a long tour of service and perhaps even makes some slight contribution to the national effort. I have always tended to count myself among this very group.

At the conclusion of my six-month course at submarine school, having been ordered to Pearl Harbor with ample delay time, I left on a leisurely camping trip in the Jeep over the Skyline Drive and Blue Ridge Parkway. For me it is second nature to head for the hills and woods, to embrace and be swallowed up by the great outdoors, forgotten by family and deep in harmony with the creatures of the wild. On the third night of my camping sojourn on Sourwood Mountain, my campsite was approached by a deputy sheriff and a forest ranger. They brought news from the Navy: if I hoped to travel to Honolulu with my family, I must be in San Francisco eight days hence. I broke camp and raced for Bat Cave, 300 miles distant.

So it was that in July 1954, I abruptly moved my resilient wife, four young children, and amiable mother-in-law from Bat Cave to San Francisco, our port of departure for Honolulu. We were to carry our vital possessions, more or less, in a tandem arrangement of Jeep and station wagon. On our last night in Bat Cave, our many good friends, grown dear with time and shared experience, gathered together to send us off with a farewell celebration at Sugar Hollow. That night, for the first time, I experienced the disarming sense of loss so often associated with change. As I looked for some assurance that our friends and this place would remain the same, I promised myself that I would always hold them unchanged in my thoughts.

Near the close of our week of cross-country travel, one hour stands out supreme. In the California Sierra, past Lake Tahoe, I spied a mountain stream too beautiful to pass and pulled to a stop off the roadside. For a short while, we played gleefully in the crystal pool, and remembered again our mountain heritage.

After a five-day ocean passage marred only by uninhibited sea sickness experienced by each member of the family in turn, we arrived off Koko Head. The rugged contours of the dank, tropic isle slid by as I pointed out to my children the landmarks familiar to me. In a matter of hours, we were settled in temporary housing on the sub base and my naval service in the Pacific began.

I was assigned duty as medical officer, Submarine Squadron 3, and, already under the hypnotic spell of deep-sea diving, took an immediate liking to my duties as instructor in the submarine escape training tank. There, in a water-filled tower 120 feet high, instructors trained all submarine personnel in a method of individual escape from a disabled submarine. The training, known as buoyant-assisted ascent, was conducted at increasing depths down to 100 feet. The trainee, wearing an expanded life jacket, would step out of a pressurized lock onto a platform in the water, where he was held by two instructors. On a signal from one of the instructors, the trainee expelled the air from his lungs until he approached residual lung volume. At that point he was released and rose to the surface at a rate of 300 feet per minute.

Throughout the ascent, the trainee continued to blow out air, since the compressed residual air in his chest could produce overdistension of the alveolar structures, immediately resulting in massive arterial air embolism. Because of this potentially fatal hazard, the submarine medical officer at all times remained in the water, working with the instructors. In case of a casualty, the doctor and victim were, within 90 seconds, moved into a topside recompression chamber.

Aside from its value as an escape training device, the tank served as a research medium in which a wealth of data on respiratory physiology was constantly being accumulated by our submarine doctors. Given the vast laboratory resources of the submarine service, the availability of tank instructors as human volunteers, and the easily controlled conditions of experimental work in the escape training tank, the horizons of research in pressure physiology were unlimited.

Tank instruction was an assignment that allowed me to dive freely and to lecture on my favorite new subjects—submarines and deep-sea diving medicine. Toward the end of my tour, I was a member of the team that made the deepest (546 feet) McCann rescue bell ascent from a sunken sub. This type of diving research became of special interest to me. Happily, I was physiologically well suited for diving and did well at it.

After a number of escapes from sunken submarines, I graduated to escaping from a sinking jet aircraft, a procedure at which I became adept, though I was never able to persuade the Navy to standardize the course for aviators. I became a living anomaly among naval officers—a doctor who couldn't fly a fighter jet, but sure as hell could get out of one if it was downed over water and started to sink. After two years of duty, with one project of independent research and intensive written examinations in the specialty, I qualified at last to wear the coveted insignia of the Undersea Medical Service—the oak leaf of the Medical Corps flanked by two dolphins.

When the final days of our stay on Oahu were at hand and it was well known that I would return to a country practice in the mountains, I had the strange feeling that something was out of sync. There I was, on the ragged edge of departure from the naval service, yet chosen as a senior technical consultant to the moguls of Hollywood for a Navy training film on submarine escape. True, I was no stranger to camera crews, having been filmed several years previously when named Doctor of the Year by the American Medical Association. (That film, which featured the planning and construction of the Valley Hospital and Clinic, depicted a "typical day" as I raced through the grinding routine of a rural mountain doctor.) And although not a career officer, I was the Navy's specialist on escape from sunken submarines and downed aircraft, so I accepted my assignment, boarded the United Airlines flight to Los Angeles, rearranged my diving gear, and enjoyed the curious stares of my fellow passengers as well as the unaccustomed luxuries of first-class travel.

Upon arrival in Los Angeles, I was met by Lt. Comdr. Carl Boyd, USN, complete with staff car and two sailors, who immediately took charge of my scuba gear, life vest, weights, and flippers. I was happy to be so warmly welcomed, and I relaxed as we proceeded to the hotel in Hollywood where I would stay until completion of the training film. Expecting to get to work immediately, I instructed the sailors to leave the diving gear in the car, as I planned to use it within the hour. As it turned out, I never did use that diving gear—or even start the training film. The whole idea was a clever pretext to ensure that I would be a thoroughly astounded guest of honor on the Ralph Edwards television show, "This Is Your Life."

At the start of the television program, I thought I had been set up for a fantastic practical joke by my friends in the submarine service. This must be their going-away gesture, satanically contrived, aimed at an old shipmate. But then came the voice of Duga Conner from Bear Branch, and in a moment she was sitting beside me on the couch, in the flesh.

The next to appear was Lonnie Hill, one of my closest friends back in the mountains, to tell the story of how we built our mountain hospital with community help. Finally, in tearful confusion, came all of my living family. Incredibly, the ruse had succeeded. In the brief time of only two weeks, the Ralph Edwards crew had pulled off a remarkable bit of chicanery. But I was grateful, happy to receive the load of gifts for me and for our community hospital, and to see once more so many old friends, though on unfamiliar shores. We all trooped off to a midnight party at the hotel where my family and friends had been hiding out for several days. When the festivities were all over, I flew back with my brood to Oahu for the closing weeks of military service.

The other guests of "This Is Your Life" went back to their scattered homes and ways of life in the Blue Ridge. At the time, Marjorie and I were concerned about Lonnie Hill, who seemed to be overly depressed about the passing of our shared best friend, Fate Heydock. Three months previously, Fate, so instrumental in construction of our hospital, had died of leukemia. I tried to talk to Lonnie about it in Los Angeles, but he put me off, saying there would be more time when we were settled back home again. Although the empathy between us was still there, a gray shade had dropped and I could not penetrate it. After he returned to Bat Cave, Lonnie sank deeper into that morass of disturbed body chemistry of mental depression. Late one night, he escaped his inner torment at the end of a rope, hanging himself from a dogwood tree near the house he had built with his own hands. He had never meant to wait for my return to Bat Cave.

In late May 1956, my reserve tour of duty terminated. I was ordered to report to Treasure Island, San Francisco, there to be permanently separated from the Navy, which I had come to love so well. Our return trip to the mainland lacked many of the cliff-hanging scenarios that had enlivened the outbound voyage. Whether the uneventfulness was due to the sophistication of the elders among us, the maturity of the youngsters, or the calmer atmosphere on board the USNS *Shanks*, I never discerned. It was a peaceful crossing. We approached San Francisco at closing speed, were brought to a secure berth, and tumbled off the ship into the arms of my young in-laws, who had driven west to meet us for a planned cross-country camping trip. My brother-in-law Roy and I, with our elder sons, descended on a local supermarket to load up for the trip. Over the next day or so, we packed my new Jeep and trailer with groceries and prepared for a seven-week camping trip, which would take us back to Bat Cave.

Return to the general practice of medicine and surgery was not so simple. The physician we had found to fill my shoes was lacking, I thought, in the characteristics I considered necessary for useful service to mountain people. Over the months, I had been alerted to this fact; however, I had assured myself and family that I could set it straight once I returned to practice in Bat Cave. I was quite wrong. As the months went by, I became increasingly aware of the irreconcilable differences of philosophy between the two of us. Perhaps it was simply that I had for so long known and loved the people on the mountain, and had always felt at home with them. Fuel must surely have been added to the fire when my former patients made fond reference to previous years of family care, and referred to Valley Clinic and Hospital as the institution they had built for Doctor George. Things like this go poorly with a freshly caught incumbent.

From my view, Bill, my successor, never felt at home with mountain people, coming as he did from a Kansas flatland heritage. As he was not trained to do surgery, he sought to refer all such cases to other hospitals and unfamiliar doctors. Quite predictably, his patients retorted that Doctor George used to take care of their surgical problems himself.

Bill's reaction was to say that I was a passable surgeon at best. That did little to improve the doctor-patient relationship, and my mountain faithfuls gradually stopped going to the hospital that had belonged to them in the first place. That was the scene I faced on my return to Bat Cave.

In an effort to keep the peace, I tried to get along with Bill but failed. His procedure of mailing bills to all patients came as an affront to many of our people, who regarded a written statement as a dun, no less. Mountain people pay what they owe when they find the dollars. Since far too many lacked the dollars, they deferred necessary care from Bill. It became obvious to me that I couldn't ease back into the picture. I realized, moreover, that I was no longer eager to return to a job that could best be described as a young man's game and, incidentally, a killer. My thoughts frequently returned to the excitement of diving research and the urgency of undertaking it. I missed diving; I missed the Navy. Before six months had passed, I decided to volunteer to return to duty.

One day in late February 1957, I received a terse communication from the Department of the Navy, ordering me to report within a week to the Medical Research Laboratory, U.S. Naval Submarine Base, New London, Connecticut. A week, hell—I could have done it in one day flat in my current state of mind! Actually, I took a full four days, driving my Jeep over the high mountain trail and camping by the roadside. There was time for reflection at night by the campfire, with snow all around in a beautifully silent world. In painful analysis those nights, I tried to justify my decision. I had brought in another doctor, our people and hospital would be served, and medical care would always be available. On my third night on the trail I convinced myself that, thanks to our hospital, the local people would probably always have competent medical care. Thus, in a sense, my original goal of establishing a hospital as a magnet for medical talent had been not too far from the mark. With that rationalization, I turned the page and moved on to a career in the Navy.

Around midnight on the last day of February, I arrived at the main gate of the submarine base in cold-weather survival gear, driving a beat-up Jeep. I was turned away despite my credentials. Next day I appeared in full uniform and was waved by the guard post without question. I was back in the Navy—back in research—and that, I guess, is where I belonged.

2

The Genesis Experiments
(1957–1962)

Our lively family settled down in a farmhouse outside the village of Mystic in the quiet Connecticut countryside, where Marjorie devoted herself to raising our four growing children. In a relatively brief period of time, I was elevated to the rank of commander and became deeply involved in the technical aspects of experimental diving research.

During the course of many months of underwater trials and adventures, I gradually assembled a team of support divers of truly superlative skills. Over the years of experimental escapes and regular sessions as training tank instructors, we developed ties of mutual dependence and deep friendship rarely encountered in any walk of life.

As a class, Navy divers tend to be remarkably clannish, and for good reason: most people regard professional deep-sea divers as a little bit daft. Individually, perhaps, a Navy diver may cut an appealing profile but, in general, we are shunned as a group. Divers, in turn, recognize the necessity for total loyalty within our small fraternity and cling together, literally, for dear life. Even so, my little coterie of fewer than a score of professionals was almost unique in our fierce loyalty. We worked together in almost perfect harmony, socialized as a team, and made up our own special squad for recovery of bodies from water under ice or within caves, as well as for unusual salvage jobs, search of submarine hulls for suspected saboteur devices, and even simulation of weightless training flights of the Project Mercury crew. In near-defiance of Navy regulations, we fraternized in good will and fellowship.

It was only natural, therefore, that as my coinvestigator, Comdr. Walter Mazzone, and I plunged deeper into the realm of physiological research, our team served as volunteer subjects for almost any experimental project in which we dared to substitute men for mice. As we explored the concept of saturation diving, I called on my courageous crew, first to assist in the animal phases as research technicians, then at last to serve as human test subjects. We deliberately set out to prove the revolutionary concept of saturation diving: Decompression time increases with depth and duration of the dive up to the point where the diver's blood and tissues have absorbed all the gas they can hold at a given depth. At this saturation point, our theory posited that the time required for decompression would remain fixed—regardless of how long the diver remained at that depth. If we could prove this to be true, the implication was that a diver could work on the ocean floor for weeks instead of minutes. In standard practice at that time, a working dive of about 200 feet was limited to about 30 minutes, as the mandatory decompression time was so long that further submergence was impractical.

But the promise of the saturation concept went far beyond the Navy's immediate needs; if valid, it would advance the premise that humans can live in and work freely from an underwater habitat placed on the ocean floor. In 1957, in one of my first conversations with Jacques Cousteau, we questioned the extended length of time required for decompression after deep dives. The only alternative that made sense was to keep the divers on the bottom until the job was done! When Cousteau learned that our group was embarked on testing my conviction that saturation diving was technically feasible, he pursued the design and fabrication of underwater habitats, while the inventor Edwin A. Link undertook the diver conveyance systems between support ship and habitat. Dr. Robert Workman, Walt Mazzone, and I continued to experiment with helium-oxygen breathing mixtures and to juggle the U.S. Navy's decompression tables accordingly.

After our six years of experimental work with animals (the Genesis A and B projects), we received permission from the Secretary of the Navy to continue the research program using human volunteers as subjects. Program Genesis C was thus begun. The four volunteers selected for the project were seasoned U.S. Navy divers, skilled in the use of underwater tools and equipment. These divers, the first U.S. Navy aquanauts, were not at that time recognized as anything special. As investigators, we weren't given priority for material or equipment; we took what we needed from any place that had it. It was rumored that base personnel locked their doors when they saw us coming.

The program began with human experiments at sea-level pressure. Throughout the days and nights of the first experiment, the physiological and psychological readings obtained from the men were seemingly unaffected by the exotic helium-oxygen atmosphere. A constant source of complaint, however, was the uncontrollable, excessive humidity, which remained near 100 percent relative.

We then concentrated on the single apparent medical problem. From the moment the normal air supply had been converted to one of helium and oxygen, it became clear that control of our subjects' body temperature was a serious problem. The temperature range within the chamber had been scheduled to 76° F to 78° F. At those temperatures, even with full clothing, the subjects began to complain of the cold in bitter Donald Duck tones. Proof of their complaints was in the shivering and occasional teeth-chattering observed at regular intervals, sometimes in unison. We had no clear picture of the comfort zone in the heliox climate. It was obvious, however, that this zone was remarkably narrow in terms of the figures established by the air conditioning industry and other investigators who had worked exclusively with nitrogen–oxygen or pure oxygen environments. A system of accurate recording of body temperatures was established that showed dramatically that, in a helium surround, the human body loses heat at a distressingly high rate. The body heat loss could be compensated for by significantly elevating the atmospheric temperature or by raising the metabolic rates of the men. The solution appeared to be a compromise between the two. The men consumed more calories, and the surrounding temperature was raised to 88° F. The thermal problem was solved, but no amount of silica gel solved the problem of high humidity, which remained a constant source of discomfort.

It was also evident that after the first few hours of exposure to the heliox mixture, the men made conscious efforts to correct the chipmunk quality of their spoken words. They did this by deliberately lowering the frequencies of the speech formats by almost a full octave and by slowing the rate of speech delivery. In addition, they learned to eliminate from their vocabulary those words that were most distorted by the helium. As a matter of passing interest, all the socially unacceptable four-letter words came through loud and clear. This prompted one of my linguist consultants to remark that, since the men were Navy aquanauts whose vocabularies were largely confined to such words, it was of little concern. I had to concede his point.

From a psychosocial point of view, a few random observations about the first experiments are worthy of record, since these patterns have been repeated in varying degrees with each successive human experiment, whether in a simulated chamber run or in later operations on the sea floor. First, it became clear that, although the authoritarian pecking order was soon established, it bore no apparent relation to the military, educational, or social status of the diver. Further, the status changed from time to time, apparently in response to emotional spontaneity and to the nature of the particular tasks the aquanauts were expected to accomplish. Yet, regardless of the changing psychosocial atmosphere within the chamber, the group of men presented a solid front at all times to Walt Mazzone and me as we stood our watches. To the budding aquanauts, we were dutiful attendants expected to satisfy any request, reasonable or otherwise. Although at the time we did not recognize the significance of the master–slave role, in later experiments it would become known as the "aquanaut breakaway phenomenon," and would become a source of potential disaster.

By the time the first critical phase of the Genesis program was successfully terminated, we had demonstrated that humans could live for several days in a nitrogen-poor, heliox atmosphere without discernible physiological damage. To Walt and me, that was a clear and beautiful ray of light, worthy of publication. That was not the judgment of our parent bureau in the Navy, however, and the experiment was not officially recorded at the time.

Seven months later, we began the next phase, Genesis D, at the U.S. Navy Experimental Diving Unit (EDU) in Washington, D.C. Essentially the same group of volunteers was used for test purposes. Now, however, the factor of pressure was imposed. The aquanauts, living in a pressure chamber immersed in a water tank (or wet pot), were exposed to the depth equivalent of 100 feet of seawater for six days while breathing a helium–oxygen mixture.

The pressure chamber had the customary "lock-out" system, whereby test samples could be passed from the inner chamber to an outer, equally pressurized space, or lock. When the inner chamber's door had then been secured, the outer lock could be opened to replace samples, remove garbage and wastes, and return large objects to the inner chamber without compromising the pressure level. The tempo of experimental investigation was accelerated. Under difficult

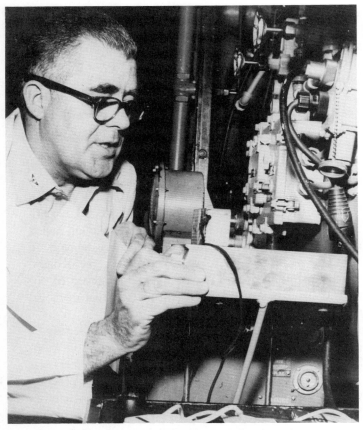

George Bond talks with the Navy's first aquanauts at the U.S. Navy Experimental Diving Unit in Washington, D.C. (U.S. Navy photo)

conditions, more than fifty physiological and psychological readings were received from each subject every 12 hours. The tests were conducted in the dry atmosphere of the chamber as well as in the water environment of the wet pot. Ambient conditions were under better control, with relative humidity maintained at about 80 percent, and such added creature comforts as we could provide.

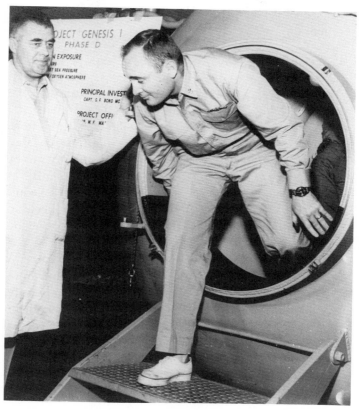

George Bond welcomes Navy diver Bob Barth as he emerges after six days in a dry pressure chamber immersed in a water tank at 100 feet. (U.S. Navy photo)

The results were excellent. Throughout the run, the "saturated" aquanauts were in good spirits. Since much of their time was spent out of the chamber in the wet pot, we judged that the simulation was approaching the true undersea environment. Blood and other biological samples were drawn at regular intervals, passed through the medical lock in the side of the inner lock of the chamber, immediately immersed in ice, and rushed to the National Naval Medical Center for rapid analysis.

Generally speaking, living conditions in Genesis D were vastly superior to those of the sea-level exposure, and the morale of the subjects improved. In a sense, this surprised us, as it was clear to the men that any significant material failure would be almost instantly fatal. Once saturated at a depth of 100 feet, there was no possibility of abrupt return to sea level. In addition, the aquanauts must have realized that they would be the first human beings in history to be decompressed after a saturation exposure. Inasmuch as Walt and I had revised the original decompression schedule many times, the prospect for success was not absolutely assured. The faith the aquanauts had in us throughout this crucial experiment proved that a vital ingredient of success in such hazardous programs is a high degree of empathy, loyalty, and trust between investigators and volunteers.

When the time came to start the actual decompression procedure, Walt and I were exhausted after 18-hour chamber watches and a fair amount of psychic strain. The ordeal of the decompression did little to improve our appearances, leaving us in limp condition to face the battery of cameramen when the aquanauts at last walked out of the chamber in perfect health. Two hours later, we took our hero aquanauts to a seafood dinner at a nearby cafe. We drove the Navy truck, parking it beside the restaurant. After a delicious meal of spiced shrimp, we came out to find our truck blockaded by miliary police, shore patrol, and Air Force police. They informed us that government vehicles could not be parked indiscriminately on city streets. Furthermore, our vehicle was from another naval district. We were all wearing dungarees and sweatshirts; some of us had not shaved recently. A military police sergeant asked to see our identification cards. We refused. We asked to talk to higher authority.

Three hours later, we were released from interrogation and began the long drive back to New London. Walt dropped me at my house long after midnight, and I sat until daybreak in the kitchen with Marjorie, sipping bourbon, discussing the experiment, and absorbing the details of our oldest daughter's elopement two days previously. I realized, not for the first time, that for months I had been far more occupied with the Genesis experiment and our aquanauts than with my family. I fervently hoped that it would not always be so.

At sunrise the phone rang. It was an admiral in Detroit, calling for advice about two men incarcerated in the forward battery compartment of a submerged submarine used for reserve training. They had been under air pressure of about 50 pounds per square inch for 48 hours. The *U.S. Navy Diving Manual* gave no hint of how they could be safely decompressed. Could I help? Thank God, I could. At 0700 I was in flight to Detroit for the emergency saturation decompression. I reminded my fatigued brain that at least I had the fresh experience of Genesis D behind me. After a hair-raising but successful recovery of the two men from the bottom of Lake Michigan, I wrote a Navy directive to fill in this lapse in the *Diving Manual* and then joined Walt to plan our next Genesis venture, appropriately named "E," as in Exodus.

With the successful conclusion of Genesis D, no slack was permitted in the program. Even as the considerable data were being processed and studied, Walt and I huddled almost daily for conferences with staff from the various scientific disciplines of the Medical Research Laboratory. With the recent installation of our new environmentally controlled compression–altitude chamber, we now had the most sophisticated chamber habitat in existence, coupled with a degree of enthusiasm that spread beyond the confines of our laboratory to envelop the entire submarine base. To a much greater extent than before, assistance in the form of men, material, and ideas came from every quarter, often unsolicited and sometimes helpful.

The date for commencement of Genesis E was targeted for August 1962. For this final laboratory experiment, we planned to expose three men: a medical officer and two enlisted chief petty officers, to a simulated depth of 200 feet for a total "bottom time" of twelve days. Insofar as possible, existence within the chamber would be autonomous. Food, water, and all other material necessities were stowed in the chamber prior to occupancy. The prepared diets, although of excellent quality and quite expensive, were packaged in drab olive containers stamped with a government stock number and, consequently, were judged unpalatable from the outset. In research, as in industry, consumer acceptance requires careful consideration.

As was the case in Genesis C and D, the artifical atmosphere of the chamber habitat would be "captive" in the sense that it would be continuously scrubbed free of excess carbon dioxide and offensive odors, while oxygen would be replenished to meet the metabolic needs of the occupants. Helium would be added from external sources to make up for normal leakage. Temperature and humidity could be programmed and automatically adjusted from the outside control console.

Portholes in the living compartment of the chamber were constructed of one-way glass to augment the sense of isolation on the part of the aquanauts and to permit continuous outside monitoring. Communications could be maintained via a two-way general broadcasting system, and a private telephone was provided for person-to-person messages. A medical lock in the side wall of the main pressure chamber provided easy access for transfer of biological samples and other relatively small objects, minimizing the gas-consuming use of the outer lock of the chamber. The entire complex had been carefully checked out by company engineers and certified as safe for human occupancy. The experiment was to begin on 1 August 1962.

Around midnight before the start-up day, the last of the internal items had been rechecked and properly stowed by our aquanauts while Walt and I completed a final check of all lines, valves, and the automatic control-recording systems installed at the dual operators' console. During the final cup of coffee for all hands prior to sack time, I gazed at the quarter-million-dollar monster to which our aquanauts would soon be committed for nearly a fortnight. Compared with the crude, hand-operated chambers that we had previously worked with, this marvel of engineering sophistication was an order-of-magnitude expansion of the state of the art. Every operation that under standard procedure would be accomplished manually was now performed automatically by a single cluster of programmed sensor–controller systems. Since these systems represented the ultimate in engineering technology, they were, of course, infallible, and almost no manual override provisions had been included in the final design. I stared at the console for a long minute; the dials, colored buttons, and strip charts stared back impersonally.

It was then that I remembered that during the previous workday, the 24-inch explosive decompression valve had been briefly disassembled to provide access to another compartment, then replaced. Certainly it must have been pressure-checked after reinstallation, but the engineering checkoff log had been carried back to the motel by the chief test engineer and was not available.

"Let's throw in an overnight pressure test for luck," I said.

We put down our coffee cups and approached the console. Walt Mazzone and I sat in our swivel chairs and began the air pressurization sequence, our aquanauts in the shadows behind us. We clicked our stopwatches together. Walt flipped *Auto Control,* and a green light appeared. Next he punched *Sensors,* then *Controls,* and once more green lights sequenced across the panel. I then dialed *Air* from my seat, and listened to the seven solenoids click into the programmed open and shut positions. "Secure main hatch!"

Outside the chamber three aquanauts easily swung shut the massive entrance hatch and gave the dog wheel three full turns. The last red light on my console turned green. All possible exhaust penetrations were secured. Elapsed time showed 28 seconds. I programmed 300 feet of depth at a 100-foot-per-minute rate of pressurization and inserted a master key, turning the switch to *Pressurize.* The last solenoid opened, and air blasted from the reservoir into the main chamber. Strip charts for pressure, temperature, and humidity rolled smoothly, reflecting the rapid environmental changes within the aquanauts' next habitat. The absolute-pressure gauge climbed steadily, touched 148.1 pounds per square inch. Air flow stopped and the room was silent.

"On the bottom," Walt said quietly. We clicked and compared stopwatches. Elasped time was 3 minutes, 42 seconds. Walt and I swiveled our chairs around to face the subjects. We grinned at one another. Then came the blast, shattering glass in the room windows and rocking the chamber itself. The chamber pressure gauge rolled swiftly counterclockwise to 14.7 pounds per square inch—sea level. The infallible chamber had failed, blown the cork! All of us in the room knew that had it been occupied as planned, three of us would have been dead. Walt and I looked carefully at our aquanauts. To a man, their faces registered disgust, but no fear. Bob Barth muttered an unprintable phrase, and Tiger Manning said, "Back to the drawing board, gents." We started a methodical inspection of the superchamber.

Within minutes we had pinpointed the reason for the calamity. The 24-inch valve, reassembled the day before, had been connected with a reversed gasket. In addition, the flange had not been properly torqued down. When the gasket blew, the metallic separation was nearly complete, allowing explosive decompression of the chamber. Our work was cut out for us.

By 0400, a huge hydraulic jack and assorted tools had been commandeered from undesignated departments of the submarine base, a 1,200-pound valve had been disconnected, the gasket turned, and the whole thing reassembled. A 1-hour pressure test at 150 pounds per square inch held without leaks. Our damage control system was reviewed. With the exception of two penetrations, all avenues of gas leak could be controlled by the chamber occupants. Both of these penetrations, the medical pass-through and the explosive decompression valve, were too large to be plugged and had simply to be trusted.

We enjoyed the luxury of a 4-hour sleep, then began Operation Genesis E promptly at 0800 hours on 1 August. For Walt and me, this would be a period of nearly continuous sleeplessness for twelve days. The aquanauts entered the inner lock of the chamber and set up housekeeping for a long 200-foot saturation dive—a first in this new diving game.

For the first 12 hours of the experiment, our gas mixture of 3.5 percent oxygen, a bit of nitrogen and argon, the rest pure helium, was easily maintained; carbon dioxide levels were controlled at our required percentages. The physiological and psychological test programs were on schedule, and Walt and I indulged in a spate of optimism. Close to midnight, the time had come to pass blood and other biological specimens through the large medical lock built into the side wall of the chamber. This lock-through was designed so that a green light showing outside indicated that the inner lock was secured, hence it was safe to bleed off pressure in the medical lock and then open the outer hatch door.

That is how the procedure should have gone, but not so. When I punched the appropriate buttons, got the green light, and undogged the outer hatch, the hatch flew out, striking me in the chest, and a deadly blast of chamber atmosphere struck my face. Squalling for help, I braced my feet and body against the brick bulkhead and forced the outer door shut. Walt rushed around the chamber and dropped in the locking pin. The crisis was over. We had lost only 10 pounds of pressure—but another 10 pounds and we would have lost the aquanauts themselves.

Walt and I resurveyed the entire engineering system of the touted chamber. I put in an emergency call for my son-in-law, Bill, a graduate engineer, and the three of us pored over blueprints and specifications. Shortly, it became obvious that the set of prints that we had approved were not represented in our chamber complex.

The nearly fatal episode of the faulty medical lock brought us all to the shattering realization that we were running an unsafe experiment. At 0500 on day 2, Walt, Bill, and I began to check each hull penetration that could present an unstoppable leak should the automatic systems fail. Within 18 hours, fifty-four of the deadly spots had been identified and suitable hand valves had been installed, replacing the automatic system.

Early in the game, Walt and I, and then the aquanauts themselves, thus lost faith in the operational integrity of the complex chamber in which they were incarcerated. It was frightening to all of us. Weeks before, I had planned to simulate a chamber casualty so that we could check the physiological and psychological response of the aquanauts to a potentially fatal situation. I could now delete that item from the program; unscheduled, potentially fatal hardware failures occurred with such regularity that simulation would have been a farce. At the conclusion of our twelve-day pressure exposure, more than two hundred equipment component failures had been detected and resolved.

Nevertheless, the experiment continued on schedule. Our aquanauts gave pints of blood and other biological products, we completed the necessary analysis, and Walt carefully recorded it all. Hour by hour we pored over the results—all acceptable by our standards. Genesis E was on the road, come hell or high pressure! Walt and I worked overlapping 17-hour duty sessions, with minimal problems. Despite the engineering failures, we had control—or so we thought.

On 8 August, day 8 of the experiment, the aquanauts, like the outside investigators, had settled into a self-satisfied air of near-boredom. Lights were dimmed in the chamber. At 0413 hours, I yawned for the tenth time and made another entry in the log. Ten seconds later, all hell broke loose. The first shock was a blast resembling a sonic boom. Clearly, we had sprung a leak and the aquanauts were in danger. The internal pressure gauge began to fall. I threw open the high-pressure air source, hit the panic button, and raced to find the source of the leak. The chamber, now overpressurized, was held at 310 feet while Walt and I assessed the situation. Gas had crept into the fluid element of our atmosphere control system and displaced the liquid in the coils, finally blowing the cap. Within a few minutes, the leak was stopped internally by our aquanauts. The emergency was over.

Next day, the completely automated control panel, designed for strictly hands-off programmed operation, began to demonstrate a dyspeptic tendency that escalated into a catastrophic bellyache for Walt and me as we watched the progressive demise of system after system. Initially, it was possible to insert jumpers in circuits, ablate some automatic controls, and continually rewrite our operational checklist. But by day 10, the careful instructions Walt left for my day watch covered seven pages, characteristically filled with admonitions such as "before crossing emergency jumper 210 to post 18, check humidity readout. If this reads 92 on the dial (subtract 8 for true value), then this jumper cross is unsafe,

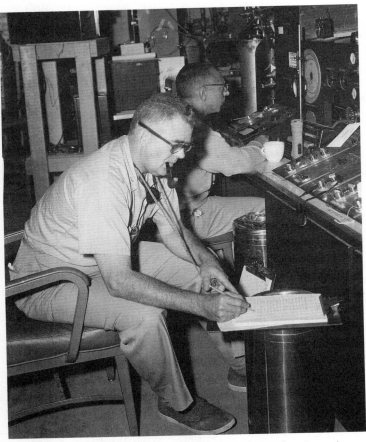

George Bond, "Papa Topside" (foreground), and coinvestigator Walt Mazzone (at controls) work out the first of the revolutionary continuous-rate-of-ascent decompressions for saturation diving at the submarine base in New London, Connecticut. (U.S. Navy photo)

and the problem must be re-evaluated." It was clear that the automated system could not be trusted. We shifted to operations by hand, after carefully labeling, with large red Out of Order tags, every one of the three-score miscreant electronic controls. The control panel and most of the chamber now looked like a decorated Christmas tree.

Inside the chamber, the aquanauts continued to perform admirably, reeling off yards of valuable data, and enlivening the action with practical jokes nicely calculated to advance the physiological age of the two investigators. "Channel fever" was the certain diagnosis, but there was also a need to vent emotionally before commencing the dangerous phase of decompression. This reaction was understandable, but as outside operators, Walt and I had lost our sense of humor.

At last it was time to commence decompression. On the basis of some scant animal work and deep personal conviction, Walt and I had agreed to attempt a schedule of continuous decompression in a gradual, continuous rise to the surface, predetermined at a rate of 12 feet per hour. This schedule calculated that a human's slowest tissue would require 180 minutes to achieve 50 percent desaturation. In retrospect, we know that this was a dangerous schedule. I must have sensed this, since I halted the continuous decompression twice, on pretext of checking all valves of our external system. It may well be that these pauses, totaling 2 hours, averted catastrophe.

During the decompression, Walt and I worked with stopwatch synchronism. The decompression was calculated for about 26 hours, with the two of us sitting side by side, save for rare departures to satisfy any urgent call of nature. Less than a day later, we completed decompressing the aquanauts. After an additional 24 hours of observation and physiological testing, Genesis E was complete. Our data indicated that we were ready for open-sea application of an idea that had germinated some six years before. Our next step: Sealab I.

3

Sealab I
(1964)

I had made several proposals to the Office of Naval Research (ONR) for an ocean trial of the concept of saturation diving, ranging from using a bottomed nuclear submarine as an ocean floor habitat, to conversion of a spherical Texas Tower escape capsule to support two or three divers for several days. I readily admit that the submarine proposal suffered from the difficulty of obtaining long and exclusive use of a high-priority operational submersible, and the Texas Tower escape capsule would be a very cramped and marginally usable device. But when ONR in fact decided to sponsor an ocean trial and we could begin planning the construction of an ocean floor habitat, there was an obvious stumbling block: we did not have the experience to determine acceptable design criteria for such a structure. As in the case of any pioneering project, the only available guide is sound engineering judgment, backed by competent counsel in good seamanship and conventional diving techniques, along with a willingness to accept a rather large number of mistakes as long as one can learn from them and prevent them from occurring where they would impose hazards to human life.

The Bureau of Ships (BuShips) has a long history in extending the Navy's diving capability. In a conference with BuShips in December 1963, bureau support was obtained, and it was tentatively decided that the U.S. Navy Mine Defense Laboratory, Panama City, Florida, would provide design and shop services for construction of the habitat to be used in the project now titled Sealab I. The swimmer research and development group of the Mine Defense Laboratory would play an important role in making competent advice readily available to the design engineers and shop personnel.

We felt that primary considerations in selecting the location for the initial run should be the prospect of good weather, good underwater visibility, a level bottom, moderate water temperature at depth, and knowledge of the general oceanography. The site adjacent to the Navy's oceanographic research tower, Argus Island, offshore Bermuda, seemed ideal; whatever the disadvantages of this site, such as the remote location, lack of swell protection, and absence of extensive support facilities, they would be outweighed by the benefits.

Approval was received from the assistant secretary of the Navy (R&D), and early in February 1964 a formal request was made for fleet assistance in carrying out the operation. In March, the Chief of Naval Operations assigned the office of the Commander, Operational Test and Evaluation Force, to coordinate fleet support. The large covered lighter *YFNB-12* was made available to serve as the principal support vessel for Sealab I.

Plans proceeded for completing the underwater structure and conducting the operation during the summer of 1964. In discussions with the Mine Defense Laboratory (MDL), we decided that some old experimental minesweeping floats would serve as a suitable shell for the main body of Sealab and that work should start immediately to fit one out as a laboratory. The work performed by the ONR project officer and the MDL engineering and shop force in readying the hardware for this project in the short space of three months is truly extraordinary—most observers would not have been surprised if completion had taken a year.

On 20 May 1964, Sealab I was ready for sea trials. The

YFNB-12 and Sealab I were towed to an offshore tower in 60 feet of water off Panama City, Florida, and Sealab was made ready for lowering to the ocean bottom—a seemingly simple exercise that burgeoned into a nightmare. Final ballasting was done the next morning and slight negative buoyancy was attained. Because of a misunderstanding in line handling orders, Sealab sank about 10 feet below the surface. As the bottom hatches were open and there was not sufficient air blowing into Sealab to keep water out, it picked up enough extra ballast to sink out of control to the bottom in 60 feet of water. While several nylon lines tending Sealab were severely stretched, none parted and, most fortunately, no one was injured. Sealab was flooded more than half-full of seawater. The habitat was blown dry of water by raising the pressure inside the habitat, and preparations were made to start lifting it from the bottom. As Sealab was ballasted to only 1,000 pounds negative, expanding air began to bubble from the bottom hatches as soon as it was lifted from the bottom. Sealab formed its own air lift and came riding up to the surface. Quick action on the part of the line handlers and winch operators prevented it from sinking a second time. Sealab was towed back into port, and MDL worked around the clock to get it ready for sea trials again. By noon on 26 May, Sealab was waterborne and trials were underway. On 27 May, Sealab was successfully lowered to the ocean floor in 60 feet of water and left for a period of 30 hours to test all systems. It was not manned. With the test completed, *YFNB-12* and Sealab returned to port, and preparations were made to get underway for Bermuda.

However, we then began to experience weeks of delays caused by weather, loading, checkouts, and loss of divers recalled to their regular assignments from their temporary additional duty (TAD) on Sealab. Then came tragedy. On the morning of 30 June, two Air Force planes were conducting a combined paramedic ocean drop exercise in making a film about Project Gemini. For reasons never determined, the pilot of one plane swerved into the path of the other. There was a violent midair collision, followed by a minor explosion, and the aircraft began to disintegrate. Together, the two craft plunged into the sea. In minutes the wreckage sank, a few bodies floated free, and the balance of personnel were lost in the sea. Military scuba divers ordered to the scene were quickly recalled because of water depth and the presence of many sharks. A few bodies were recovered by small-boat crews, but at least a dozen paramedics and crew were missing and presumed dead.

Later that day, when I heard of the tragedy, I volunteered the services of our topside diving crew. A few hours later, we were informed that water depth at the site of the sinking was about 720 feet, well beyond the limits of any conventional diving. We proceeded with our planned schedule of Sealab operations. Next morning, however, a reported positive fix by the Coast Guard showed a water depth of 210 feet—too deep for standard scuba diving but within the capabilities of my group. Our services were requested for body recovery, and we were underway with the *YFNB-12* at once.

On the scene, we began a bottom search, first with divers to determine visibility and then with our underwater television equipment. It was immediately clear that the location marked by the Coast Guard was a false contact, but soon we were skirting a large area of wreckage randomly scattered at depths of 190 to 240 feet. There were very few large pieces and nothing easily identifiable. We buoyed off the more promising areas, and our divers entered the shark-infested water.

After two long days of extremely hazardous recovery operations, we encountered conclusive evidence that little in the way of human remains was left to recover. I returned to Kindley Air Force Base and advised the commanding officer that further diving search would be futile. Our assistance with the operation was terminated, with sincere expressions of gratitude for our efforts. Underwater recovery of human remains is an unhappy and unrewarding chore. In all honesty, most of us feel that, save in the case of a simple drowning in shallow cold water, it is infinitely kinder to consign the remains to the ocean depths with an appropriate ceremony. In this way, the dignity of death is preserved for survivors and deceased as well. Our Sealab aquanauts returned to Naval Station, Bermuda, and tried to pick up the lost threads.

By Independence Day, the schedule began to firm up. All preparations inside Sealab were reasonably complete and all our divers on TAD were on hand and accounted for. On board *YFNB-12,* technicians were tripping over one another in the control trailer, making last-minute adjustments on the myriad of complex electronic gear that would monitor every heartbeat and breath of the aquanauts. In truth, Walt and I were being instrumented out of working space, but we were both thankful to have such a dazzling array of the best possible gear. In this open-sea experiment, we were intent on getting a maximum amount of useful scientific data, without which real progress in undersea experimentation cannot be expected. For all the fanfare about Cousteau's remarkable 1964 Conshelf experiment in the Red Sea, it obtained not one iota of useful physiological information. Ed Link, with his electronic expertise, had demonstrated that men could live for a 48-hour period as deep as 432 feet, but no significant data were derived from his divers' exposure. The feats of both these pioneers lacked proper scientific documentation. In Sealab I, we were determined not to perpetuate such mistakes.

The next day a dispatch arrived saying that the investigating board of the Air Force crash had reconsidered its decision and now desired recovery of all aircraft pieces to determine the true cause of the disaster. Furthermore, we were requested to comment on the opportunity to move the Sealab I operation to the scene of the accident and assign the aquanauts the entire job of undersea mapping, bottom photography of the whole mess, and salvage of all parts.

At first, the opportunity was attractive. We could show flexibility in moving from a planned to a totally unplanned underwater project. We would be working on a significant operational job and proving a point I had been preaching for years—that the Sealab concept was the most effective means of doing useful underwater salvage and extensive survey. A successful completion of the task would bring tremendous popular response. The aquanauts would feel usefully employed and not like just experimental subjects for another in a long series of scientific reports.

All of these factors I considered over four pipesful of tobacco—and I rejected all of them. After Sealab was conceived in 1957, each slow step toward completion had been aimed at our first undersea manned exposure. As we approached this goal, I had selected a site of operation. The site had been studied from the point of view of five-year bottom current reports, daily sea states, uniform ocean bottom depth, and minimal obstruction hazards to the aquanaut on a sortie. To change our scene of operation and primary mission overnight might defeat the purpose of the initial experiment. At least six precious days would be lost, moving us close to the hurricane season. Important physiological studies would have to be curtailed or deleted. The ocean bottom depths were irregular, with abrupt shelving to 600 feet near the wreck site, and a new factor of decompression problems could be foreseen with no means of treatment should the divers experience bends. In addition, a four-point moor for the *YFNB-12* in this location would be technically difficult and of uncertain holding capacity, and no records of bottom currents or daily sea states were available.

After discussion with Lt. Comdr. Roy Lanphear, project

manager, we made a conference call to ONR and received a sympathetic ear. Accordingly, we sent out a joint dispatch that recommended that we stay on steady course as previously approved at all levels. The flap was turned over to higher authority, as is the custom. Unless otherwise directed, we planned to proceed to Argus Island the next afternoon.

At this time, a phone call from the station dispensary informed me that Comdr. Scott Carpenter had just been admitted following a motorbike accident, with a compound fracture of the left arm and assorted other injuries. Commander Carpenter, pilot of Aurora 7 spacecraft through three Earth orbits in 1962, was on loan to us from the National Aeronautics and Space Administration (NASA). When I saw Scott at the dispensary, I realized that his terribly painful injuries were nothing to him in view of the anguish of missing Sealab I. But after examining the X-rays, I had to tell him that he was on the scratch sheet. Off he went by ambulance for specialist attention at Kindley AFB hospital, where I revisited him a few hours later by helicopter. My guesses were right: he must be evacuated stateside. I promised him a permanent billet with our next mission.

My aquanauts, including Scott Carpenter, are truly a breed apart. They are immensely resourceful, incessantly curious, impervious to hazard, and impatient with the progress of the world. They are the ones who take the chances for all humanity, who care nothing for fame but seek only the satisfaction of battling odds and attacking new frontiers. Scott had long been in this fraternity, and his setback was a blow to all of us who had grown so fond of the man and wanted him on this first run.

The weather in Bermuda turned unpleasant. A wind of almost gale force boxed the compass with monotonous regularity; rain fell intermittently and the sea state was miserable. We decided that Sealab I could not be satisfactorily lowered by the barge *YFNB-12* with the equipment then available. To proceed with the project, we adopted a method of handling from the crane on Argus Island, and on 18 July, Sealab was again ready to go.

In the course of any underwater venture similar to Sealab I, three written reports are prepared: (1) a situation report, or SitRep, which is a dispatch of a few words to Washington, D.C., authorities, outlining all project successes and minimizing unhappy events; (2) the daily log of the operation, an expanded version of the daily SitRep, with a small amount of personal commentary by the watch-stander; (3) the Chronicle, of different caliber and texture, in which the principal investigator records a distinctly personal evaluation of the total scenario. The Chronicle is a biased account, twisted to meet the needs of the author, but it deserves some credence. For this reason, the following selections from my Sealab I Chronicle are presented without abridgement or apology, starting with the factual account of the precarious descent of Sealab I to the bottom of the ocean.

19 July 1964

Today the ocean was nearly calm and the wind slackening. We had punctured one rubber boat, and two other supporting small craft had been put out of action. But the umbilical cable had been installed at the unbelievable depth of 63 feet without wetting a single component. Our planned communications network had given up the ghost, and we were running this tremendously complex procedure with the aid of semaphore, bullhorn, runners, swimmer–messengers, and a rare bit of mental telepathy. When a communications system of sorts was established, messages went invariably to the wrong stations, and were garbled in nearly every instance. Meanwhile, Sealab I was being lowered under orders from the command at the tower, while we sat in the control shack in the trailer, desperately trying to keep up a gas flow adequate to prevent flooding. On occasion, we fell behind, or at least our gauges said so, and at least twice we had minor casualties requiring temporary interruption of gas flow. Yet somehow we got her down safe and dry—more a tribute to the built-in engineering safeguards of the habitat than to our handling procedures, which were atrocious.

By early afternoon, Sealab I was settled on the ocean floor at 193 feet, dry and with all systems functional. I breathed, with all the rest, a prayer of thanks.

20 July 1964

The day dawned clear and glassy calm. To me and to all the rest came the strong conviction that the project would go as planned. No room now for pessimism; no time for aught but to get our aquanauts into their habitat safely. We ate an early and hearty breakfast, and began final diving activities.

Our first set of divers entered the water to repeat the inspection dive and to complete unbuttoning of the habitat. They returned to report a low water level in the entrance hatches, excellent position of the habitat, and all ready for final ballasting, which began at once. One at a time, with ponderous care, the great anchors were lowered to Sealab I, delicately poised above the bins through TV control, and placed in position, after which the divers descended 190 feet to unshackle and allow the cycle to be repeated. It was slow, dangerous, delicate work.

Meanwhile, word came that the first visit of reporters and photographers was scheduled around noon for a brief tour to observe final checkout, to interview the aquanauts, and to witness their final water entry. Walt Mazzone and I, as plank owners in this long series of experiments, reserved the right of final inspection of the habitat. Accordingly, we swam from the barge to Argus Island in company with some large but friendly barracuda and entered the submersible decompression chamber (SDC).

With Chief Ray Lavoie as operator, we lowered rapidly to the vicinity of Sealab I, which soon became visible with startling clarity on the ocean floor beneath us. Remarkably, even at this late hour and great depth, the international orange, or fluorescent red, color of the habitat showed clearly. Such clarity is contrary to divers' experience and to known laws of physics, but there it was. The reflected light from the coral sand on which our baby rested was remarkably bright, and the water absolutely clear. The SDC came to a halt at about 160 feet of depth, which meant a long breath-holding swim of about 90 feet before we reached the shark cage and entered the hatch to the habitat. I went first, after a last deep breath from the area of the SDC, with Walt right behind.

The swim was beautiful, but at first a little frightening. It was certainly the deepest breath-holding skin dive I'd ever made, and possibly the deepest in history. Sealab I seemed a long way off, with much interesting marine life in between. I wondered if the back porch might be accidentally shut, or if the shark cage might be full of large groupers, as we had seen on TV the night before. I wondered if the hatches were really open, or if my divers had played a grisly practical joke on us and removed all bolts save one. The wondering ceased: I was in the cage, past the first hatch and rolling on my back to enter the hatch to the living quarters, which gaped open. I bobbed up inside, breathed a wonderful atmosphere, and was immediately joined by Walt. We announced our safe arrival in helium chipmunk voices that could not be understood topside. A fast inspection of both compartments, and we were off for a leisurely free-ascending swim back to the SDC, blowing bubbles all the way. We surfaced, swam to the barge and bade farewell to our aquanauts as they prepared to descend to the ocean floor. As they submerged in the SDC, Walt and I took our stations in Sealab control, he at gas monitors and I in communications. Minutes passed, and more minutes. Then a chipmunk in Sealab I began to sing "O Sole Mio." It was Anderson, the happy gunner's mate.

In the vertical mode, the Sealab I submersible decompression chamber transported aquanauts under pressure to and from the underwater habitat. In a horizontal orientation, it served as the on-deck decompression chamber. (U.S. Navy photo)

The Navy's Sealab I is a 40-foot undersea laboratory now on display at the Museum of Man in the Sea, Panama City Beach, Florida. (U.S. Navy photo)

Five minutes later, the aquanaut muster was complete and all hands at work. Operation Sealab I had become a reality.

21 July 1964

Walt and I are heel and toe on the watches; Walt takes the night and I the day shift, with each of us easily slipping back into the routine of earlier years. On the ocean bottom, however, our aquanauts have set a different watch. For them, with exception of the single wake man, the day commences at 0900, with activities continuing until nearly midnight. Since most of the interesting and previously unobserved marine activity occurs from sunset until dawn, these hours are tremendously productive for careful observation and photography. Late last night, Manning observed a 400-pound tuna feeding on schools of jacks just outside the Sealab. That is probably one of the very few times such a scene has been witnessed by a human, and it is only one of many unusual situations that are almost commonplace to our undersea observers.

Meanwhile, the aquanauts live, work, and swim, all at a leisurely pace that, in the course of nature, will quicken tomorrow and again the next day, until they attain a normal pace on about the fifth day on the bottom. Such an involuntary initial slowdown has been our experience in all past chamber experiments, and such was Cousteau's experience with his two undersea projects. None of us has a satisfactory explanation for the phenomenon. Perhaps it is a natural conservation of body energy, pending an unexpected environmental emergency, or awaiting the process of acclimatization to a new environment of multiple stresses. Characteristically, most mammals move slowly in the face of a potential but undefined threat, including humans in an unnatural surround.

Although this experiment is only 72 hours old, a tremendous fallout of engineering and psychological problems is evident, in addition to our mass of scheduled daily data. Thus, as we had hoped, this exercise will uncover new problems in number far greater than the questions answered. From an engineering point of view, the outstanding problems have had to do with handling a mass such as Sealab I from a surface vessel or even from an ocean platform as stable as Argus Island. Of more interest is the fact that the absorptive properties of seawater can be made to work for underwater architects in heating or cooling their houses, and in providing and purifying the atmosphere. Medically, we have found that, once the human body is essentially purged of its nitrogen content, the narcotic effect of high levels of nitro-

Navy diver Bob Barth tests an electrically heated, lightweight wet suit. (U.S. Navy photo)

gen is markedly increased. This phenomenon has constituted a serious hazard in the present experiment, where the men have been obliged to work for short periods in compressed air at the relatively shallow depth of 187 feet. Unexpectedly, after 24 hours of undersea existence, the aquanauts became severely narcotized upon entering the air space. So much was this a threat that it became necessary today to dilute the air in this space equally with helium, and to do likewise with the air in the open-circuit scuba bottles used by the aquanauts.

Incidentally, I am again a papa. Back in the days of Genesis, I was Papa Gen and my subjects Genitalia. Here I am Papa Topside. It is strangely comforting to sit in Sealab control, listening to our pioneers and watching them around the clock, feeling that they are as safe as seven years of hard work and planning could make them, and knowing, too, that we have only turned the first page of a potentially great chapter in human achievement.

27 July 1964

All things considered, it is time to terminate the bottom stay of the aquanauts and to commence decompression, providing weather conditions are nearly ideal. A good many factors enter into this decision, but the one of overriding importance is weather, as each passing day drives us a bit deeper into the hazardous hurricane season. Last year, Hurricane Arlene hit here with full force on 5 August, bringing 70-foot waves to Argus Island and damaging much of the structure. More important, the area had only 3 hours' warning of the direct hit—far too little time for protective measures for our aquanauts. While Sealab I would undoubtedly ride out such a blow very nicely on the bottom, our surface support, on both the barge and Argus, would probably be eliminated for days, if not weeks. Moreover, by compressing our schedule, we will soon have acquired all the physiological data that we can process with any degree of success. Extension of time to the planned twenty-one days would lengthen our curves, but would not likely produce much more of value. A very long stay of at least sixty days would be required to give results for a "chronic" experiment, and this we cannot do at this time. Although we have priority to dislocate other important Navy projects for our requirements, the wisdom of doing so is questionable. In the years to come, successive Sealab projects will have to live with other Navy bureaus, and it would be unwise to create friction on the first operational venture.

29 July 1964

On this day, unfavorable weather predictions prompted termination of the experiment, and at midnight Sealab was lifted off the bottom. The rate of ascent was 1 foot every 20 minutes and was uneventful until about 110 feet, when wave forces began to make themselves felt again. We stopped the ascent at 100 feet for several hours, hoping for abatement of the seas, and finally started it again. Sealab was brought to 81 feet, at which point shock loads of 20 tons were being put on the crane, and further ascent was not considered safe. The aquanauts were transferred to the submersible decompression chamber and brought up on Argus Island. Sealab was floated off on the same four buoys that had been used to transfer it to Argus Island and was brought astern of the *YFNB-12.* Ballast weights were jettisoned until positive buoyancy was attained and Sealab floated. Once on the surface, it was secured for sea and towed back to port.

1 August 1964

Like all natural phenomena, any human experiment, after its conclusion, continues to send out backwash vibrations that decrease exponentially, as with the ripple pattern on a millpond surface. So it is with human physiological and emotional circadian wave patterns after a prolonged period of artificial living and unrelenting stress. For the aquanauts and topside scientific observers alike, the days following the successful completion of the project had slowly receding patterns of activity and rest; of mental activity and emotional boredom; of hunger for, and satiation with, the company of fellow men; and finally, of happiness and marked irritability. Personally, I had an overwhelming desire for solitude and the opportunity for introspection. I sensed that the others felt the same. None of us, however, had much opportunity for privacy, the requests of the news media being what they were. So we were all a bit short-tempered and impatient with the multiple intrusions on our brief hours of solitude.

For my part, I had a great deal of writing to complete, and each interruption broke the continuity of my thought. To me, there was a great sense of urgency about making a written record of Sealab I before my impressions became flattened by the passage of time. For Walt Mazzone, the best outlet was to assume responsibility for packing and shipment of all Sealab gear destined for return to New London and for arranging return travel details for the project personnel. When he met with postexperiment apathy and he, too, became irritable, a degree of alienation occurred between investigators and subjects. But we are all bound to a common cause and we like one another very much. For a few days after any experiment such as Sealab, we should each look in a different direction until the feelings pass.

4 August 1964

Today a planeload of admirals and captains was to arrive from Washington, D.C., with journalists from all the major newspapers. Extensive preparations had been made for this session, the climax of all the Sealab I press conferences. The stage would be set in the outdoor theater and martial music would pitch the emotional tenor of the occasion. Before dawn, I was awakened by peals of thunder and blinding lightning flashes. An unpredicted squall had struck Bermuda, accompanied by lashing rains. By breakfast the rain had abated, but not for long. By the time the plane from the capital was due, torrential rains were falling and visibility was zero. We switched our press conference site to a spacious air hangar near the heliport, where a shaky platform was installed and a public address system hooked up. I tested the microphones in the reverberant hangar we had chosen and found that the echoes produced were of quality equal to those I had enjoyed as a student on the campus of my old military school, Posey, located on the Alpine slopes of Zweisimmen, Switzerland. On the whole, the sensation was not unlike being in a giant echo chamber. Strangely, the press seemed content, perhaps because if they missed an answer on the first bounce, at least a score more opportunities were presented before the words died away in a simple roar.

The questions posed were sensible, and the responses relaxed and informative—at least, up to the point at which Andy Anderson, gunner's mate, took the microphone. Andy has a fantastic vocabulary, not one word of which is acceptable in polite society. When he was called to speak, many of us felt a real need to be elsewhere. Yet talk he did, and without so much as a whisper of blasphemy, profanity, or vulgarity. As we watched in awe, Andy sweated each word, visibly sorting through his repertoire in search of acceptable rhetoric and syntax. He pulled it off, though his pauses tended to accentuate all the scurrilous four-letter words that he detoured. At last it was over, but the strain on the man was obvious; possibly this was his longest printable communication since he was weaned.

Finally, the visitors' buses cranked up and we were alone. I entered the bar at the Officers' Club wearing my hat, and was charged with a round of drinks according to naval custom. Somewhat later, I bought a bottle of sparkling wine to romance my wife on her imminent arrival and returned to bachelor officers' quarters (BOQ), where I lay down and slept serenely for 16 hours.

4

Sealab II
(August 1964–August 1965)

Sealab I was over. Our data were complete. It was clear that prolonged manned undersea ventures were practicable and relatively safe, perhaps at much greater depths. The job was done, the reports were written, but the future was bleak. We were only a handful of determined pioneers, somewhat shaken by our inability to handle the mass of Sealab in the open sea, completely out of money, and facing a bureaucracy that would have been content for us to return to matters of operational importance. The Office of Naval Research had, with difficulty, absorbed the financial brunt of Sealab I, but there was no budgetary provision for extension of the project—nor were there other decent prospects for future funding. The Bureau of Medicine and Surgery, although lately cheering from the sidelines, had no intention of picking up the tab for further work, and the tab, by my estimate, would be very large.

Back once more in New London, my assessment of the situation was realistic: we had accomplished an exceptional exercise, but we would not have the financing to extend our diving and working depth to the average depth of the continental shelf—about 600 feet.

But three of us could not drop the torch. Bob Workman continued his work at the Experimental Diving Unit; Walt Mazzone applied his newly acquired physiological knowledge to our lab program; and I worked alternately as boss of the outfit and doctor–instructor in the submarine escape training tank—back again to my original training and first love. Sealab II seemed far over the rainbow. Then I learned that public concern over our apparent inability to locate the remains of a sunken nuclear attack submarine had dramatized the need for the Navy to develop diving and operational competence in deep water. Heartbreaking as the loss of the USS *Thresher* was for the families and friends of the men who went down with her, the lessons learned from this tragedy led to the essential Man-in-the-Sea Program for the U.S. Navy.

My role in this program began unexpectedly in August 1964, a month after the wrap-up of Sealab I. I was in the Washington, D.C., area to accompany my mother-in-law to Bethesda Naval Hospital for major surgery and to remain for several days until her successful recovery was assured. While I was in Bethesda, in came a call from my old friend and commanding officer, Rear Adm. Walter Welham. Would I attend a conference with him in the office of Rear Adm. "Pete" Galantin? My answer, of course, was yes.

Admiral Galantin's workshop was spacious and quite worthy of occupancy by the director of Special Projects. After introductions by Walt Welham, Admiral Galantin ordered a round of coffee, made a pleasant comment about the Sealab project, then dropped the bomb: "George, the DSSRG [Deep Submergence Systems Review Group] recommendations have been approved and the whole program awarded to us here in Special Projects. I want you to direct the medical end of the Man-in-the-Sea effort." With boggled intellect and swirling senses, I gasped out a quick acceptance. Almost immediately, plans were put in motion to design a new, improved habitat and to select and train more than a score of aquanauts.

Sealab II would be an underwater house capable of sheltering ten aquanauts at a depth of about 200 feet for a period of thirty days. The habitat's optimum size would be 50 feet long and 12 feet in diameter, with four separate areas: an entry, laboratory, galley, and the living space. As the atmosphere would be predominantly helium, extra insulation would have to be provided. Temperature would be held at 88° F and humidity at 60 percent relative. The Hunter's Point Division of the San Francisco Bay Naval Shipyard would design, construct, and outfit this new marvel of the undersea world.

The selection of Hunter's Point was based on the yard's experience in building submarines and its willingness to undertake the job for completion in the specified time of four months, which would be nothing short of a miracle. Lt. Comdr. Malcolm MacKinnon III, USN, assistant planning and estimating superintendent for submarines, took charge. It became apparent to him early on that this operation was going to proceed under a different set of ground rules than those he was accustomed to. Clear-cut specifications and contract plans were nonexistent. Basic design decisions were made and production was often started from as little as a pencil sketch scrawled on a paper napkin during lunchtime. He frankly described the project as "a bootstrap operation all the way" and brought the full range of his engineering expertise and considerable common sense to designing and fabricating the vessel from our wish list.

Right at the outset, MacKinnon conceded that the nature of the Sealab project, primarily research and development involving a large number of people and activities, made design of the habitat a nightmare in systems engineering, in that each decision either required interaction with, or was governed by, other members of the project team. In addition, the guidelines we advanced to him, based on feedback from the difficulties encountered in Sealab I, were unique in terms of provisions for personnel safety and physical comfort, which he held paramount throughout the design process.

As a result of the flooding that we had experienced more

than once in trying to lower Sealab I while simultaneously keeping internal gas pressure slightly higher than hydrostatic, we asked that the new habitat be capable of being fully pressurized to bottom pressure *prior* to submergence. MacKinnon noted that, with this requirement for full working pressure within the habitat while still on the surface, Sealab II had to be an internally nonfired pressure vessel under the American Society of Mechanical Engineers boiler code. The cylindrical shape, moreover, required an ellipsoidal dished head at each end, and the threat of a national steel strike meant that the ellipsoidal dished heads were not available commercially within the short delivery time of thirty to forty-five days. The shipyard thus entered a new field of endeavor: the underwater explosive forming of large steel sections. Fortunately, however, several small-scale steel pieces had been successfully formed using this process at the Navy's West Coast Shock Testing Facility, conveniently located within the shipyard. Meeting the challenge head-on, MacKinnon decided to make a quantum jump from the small, simple pieces previously fabricated to a 12-foot-diameter, 1-inch-thick, mild steel section with complex curvature. Happily, the results were phenomenally good and only minor straightening in certain areas was required.

To eliminate the disasters encountered with Sealab I in the submerging phase, MacKinnon designed the habitat along submarine principles. The variable ballast would be water, providing stability during all phases of submergence, as well as negative ballast on the bottom to ensure firm seating. The Naval Ordnance Testing Station in Pasadena developed an ingenious winch-counterweight lowering system to make feasible the controlled lowering of Sealab II against negative buoyancy. The necessary negative buoyancy would be obtained by flooding tanks 1 and 3 and the conning tower. A remote-reading level indicator measuring pitch angle and list would be monitored during descent to effect proper bottom positioning. When the bottom position was deemed satisfactory, tank 2 would be flooded by surface divers, making Sealab II ready for our aquanauts.

To raise the habitat, the resting pads would be washed clean to break any suction and tank No. 2 blown dry. The lowering process would then be reversed, hauling up the 5-ton negative buoyancy. When the top of the conning tower appeared at the surface, the conning tower would be blown dry and Sealab II would float at the midheight of the conning tower. The pressure would then be released from inside the habitat, blowing tanks 1 and 3, and Sealab would float at 1 foot, 8 inches freeboard, her normal surface condition.

The end result of Lt. Comdr. MacKinnon's efforts was a magnificent piece of engineering. Almost precisely one year after completion of the Sealab I experiment, we had a brand-new habitat, to be placed at 205 feet on the edge of Scripps Canyon, several hundred yards offshore La Jolla, California. In contrast to the warm, clear waters of Bermuda, our undersea environment was cold, dark, and damned hazardous. Our primary source of surface support was an improvised catamaran originally configured to assist in the initial Polaris missile underwater launchings off San Clemente Island. Many modifications were made in the vessel, including stowage for the personnel transfer capsule (PTC). It was the means of returning the aquanauts to the surface. The men entered the PTC and, by closing the bottom hatch, trapped the ambient bottom atmosphere and pressure. The ascent was then made using a winch on the staging vessel. At the surface, the capsule was transferred to a crane, which lifted it clear of the water and into position for mating to the deck decompression chamber (DDC). The pressure was then equalized between the PTC and the DDC, and personnel were transferred to the larger DDC for decompression. The PTC could also be used in an emergency if the habitat had to be vacated; it was kept on the bottom with a clump of lead weights. When the weights were cast off, the PTC, under positive buoyancy, could be raised and restrained by a rachet mechanism; a cable was firmly attached to the ballast clump. If necessary, decompression could take place inside the PTC on the surface. A small diving bell was provided for short-term excursion dives to the habitat by surface support divers. In addition to the staging vessel, the salvage ship USNS *Gear* and various service craft would be available if needed.

By July 1965, I had at last finished with the travel, conferences, and desk work associated with the planning of Sealab II. I joined the project members on station at the Long Beach Navy Yard, California—the finest group of men who could be gathered for this project. We stood nearly four hundred confident people, funded to more than twenty times the extent of Sealab I, and with the interest of our nation and our Navy supporting us. It struck me that the whole story of Sealab had been a tale of difficult escalation against seemingly hopeless odds. My first proposal had been rejected with the curt official remark that it was ill-conceived and impossible to accomplish. I was admonished to work on matters of operational importance. At last, eight years later, we were underway.

10 July 1965

Things are moving along at a good clip, with the staging vessel getting a badly needed coat of paint, the audiovisual trailer control being outfitted, and equipment rolling in from every direction. After a new suit of paint, the staging vessel, renamed the *Berkone*, after the combined names of Joe *Berk*ich and Walt Mazz*one*, will look as good as any staging vessel can. Walt and I are pleased with our accommodations topside, where the decor reminds me of *Voyage to the Bottom of the Sea*.

This has also been a time of waiting, retreat, and self-assessment by everyone connected with the program. A few weeks ago, in the depths of North Carolina's Linville Gorge and Gingercake Gap, I had the opportunity to study the stars, the universe, and my own soul in solitude. I find that each one of us in the program goes through his own reach of self-examination. In many cases this scrutiny is postponed by the excuse of late, wild hours of play or by impulsive, prolonged hours of work. But each one of us, it is hoped, will find a time to square this enterprise with himself and his God.

Shortly after my arrival at Long Beach, Capt. Jack Dolan came by for a fast walk through Sealab II and a brief conversation. Like myself, he is dumbfounded by the escalating cost of the project, and rather articulate on this point, undoubtedly a sore one throughout the Deep Submergence Systems Program (DSSP). To me, the whole matter of cost estimates and program management is a closed book and probably one that I will never open with any degree of success. To me, it is fun to run a program, but I find managing one to be a tricky system of double-entry bookkeeping. A financial manager does not bear responsibility for the end result. Unlike the investigator, the financial manager is not expected to identify with the emotional demands and physical rigors of the divers and support people. I find it impossible not to have empathy with the doers, and it shows in my management.

16 July 1965

This day was not improved by the sad spectacle of Adlai Stevenson's funeral services in our nation's capital. I couldn't help but speculate that had the nation seen him in action as a public servant as in his later years, rather than as a campaigner, we would have better measured the worth of the man and he would have been a runaway favorite for president. Adlai Stevenson served his country and mankind in a superlative fashion, and his mark in history is assured.

Now come the people problems that have to be anticipated in the latter stages of a project. From every quarter I am told of apparent conflicts of management and control of vital program elements, team personality conflicts, and alarms and loud noises all over. Much of this I have provided for, yet a few people in the program feel a compulsive need to uncover new (to them) problems, often only to underline their own personal value in the program. For this very reason, I established clear-cut lines and areas of responsibility months ago. This day alone, I received several false reports, any one of which, if true, could have stalled the project. All were incorrect; more important, not one was checked out before being brought to my attention. Tomorrow, the tempo will again increase. After next week, we are locked in, with the job in its final phases of acceleration. No more time, then, for reflections or second guesses.

20 July 1965

A call from Capt. Jack Bennett concerned his efforts to change our billets (assignments to duty) from temporary additional duty to permanent assignment for the duration of the project. Jack worked hard on our behalf just before his transfer, and he was assured that our billets would be immediately forthcoming. Alas! When his back was turned, so were the tables, with a prompt return to status quo. I plan to visit briefly with him in San Diego, and to have Scott Carpenter join me in a presentation to a group of young officers at Ballast Point. Scott, happily recovered from the motorbike injury that prevented him from taking part in Sealab I, will serve as a team leader for thirty days in Sealab II. Scott is the first American aquanaut–astronaut and one of very few people I know with the required charisma, knowledge, and experience to endure the tests he will undergo during his month at the bottom of the sea.

Today, more news that the Bureau of Medicine experts will require added paperwork relative to my crew selection and other items. The hour grows late, and shuffling papers at this point could do real damage to the program, but I make the effort to comply because I can do nothing better in the interest of the project. These late-season willies are also apparent in the Special Projects Office, where I will be required to justify at length a number of critical program features I had considered to be accepted fact. I sense a great deal of BuShips influence in these negative developments and predict that considerable effort will be expended to prevent our use of the MK I swimmer propulsion unit. I'm keeping my fingers crossed for a favorable hearing.

It is because of the ultraconservative, nilhilistic attitude of many people in the Navy that diving practices and capabilities made no appreciable progress during the past quarter century. Now comes our operation, and backsliders set into it with a will, perhaps because it shows up their own unimaginative programs. But we have made progress and will continue to do so. Our project has too much intrinsic merit to suffer defeat because of petty personal jealousies or bureaucratic politics. Still, it would be an easier and happier road if we didn't have to fight every step of the way.

23 July 1965

Today Sealab II was christened. At first blush, things seemed to be in reasonably good order. That in itself is a bad omen in any of my projects. It is the simple, straightforward situation that will inevitably parlay into tragedy or worse. The morning press conference was scheduled in a working area of the shipyard directly in front of the impressive mass of Sealab II. A battery of microphones was set up for all major networks, and TV cameras were appropriately placed. All hands were properly garbed, and the setting seemed ideal. At this point, the situation began to deteriorate.

Commanding officers on two of the nearby Navy fighting ships made a decision to test their radar equipment. Ordinarily, these test sweeps are innocuous enough, but on this day they were unerringly tuned to the frequency of our public address system. The result of this combination was a series of waves of nerve-shattering sound sweeping the area at 15-second intervals. Angry protests were relayed to the commanding officers of the offending vessels, but without results, save for a few clearly audible and unprintable retorts asserting their prerogatives to test equipment. The news conference continued, with each speaker trying to time the words between decibel blasts and being consistently outguessed by radar operators, who changed their timing on each sweep.

Ultimately, quiet was restored and a deadly silence blanketed the scene, punctuated only by two wildcat chippers who had found a piece of steel to hammer. Capt. Lew Melson presented a general rundown of our program. Speaking without notes, he started off splendidly, but later the factual material assumed a life of its own in the media. I heard that the operation would take place *in* the Scripps Canyon, that the habitat would be located at 215 feet (we had slipped 10 feet downhill), that Sealab II had gained 25 tons since yesterday, and that we would provide all gas requirements from the staging vessel. (This may not be an exact record of the apparent errors, but the facts came out distorted that way in the news.)

The christening ceremony itself came off quite well. Capt. Jamie Adair led the show with exceptional ease and grace. The parade of the colors was moving and the benediction more so, as a splendid band played a muted Navy hymn throughout. I wept unashamed and not alone. Next, a beautifully prepared speech by the Secretary of the Navy, a smashing blow on the nose of Sealab II by Heidi Nitze, a concluding prayer, the retiring of the colors, and the show was over. Our aquanauts and the supporting team were magnificent throughout a great day.

24 July 1965

Our Sealab II crew is largely handpicked and, in my opinion, the most highly motivated and best trained group available in the U.S. Navy. However, they possess complex personalities and are prone to development of prima donna attitudes with minimal provocation. To an extent this is healthy, since they must see themselves as a breed apart if they are to perform at maximum efficiency.

Since the tragic chamber fire at the Experimental Diving Unit last spring, which brought instant death to two of the Navy's finest divers, one of the aquanauts has shown a marked recurrence of outbreaks of his latent rebellious attitude against authority. This is not a new attitude for A, as I will call him, but I believed that he had whipped himself into line and would fit into our command structure without further outbreaks. I was dead wrong. After we arrived at Long Beach, and without any advance warning, A chose to challenge authority. The instrument was simple—as uncomplicated as the reasoning of the aquanaut: A refused to participate in morning physical training. When directly ordered by Comdr. Scott Carpenter to join in, he disobeyed the order. When Scott attempted to reason but got nowhere, he disqualified A from the aquanaut team.

For reasons apparent only to himself, A thought that I would take his part and overrule Commander Carpenter. Further, he informed Scott of this belief as an added goad to authority. I was told of the situation, reassured Scott that I agreed with his decision, then sought to get hold of A, without success. He evaded me throughout the night, and was absent during the christening of the habitat yesterday.

At long last, this afternoon he became available, thanks to Navy divers Bob Barth and Wilbur Eaton. A and I had a closed session at the motel. I pointed out the obvious fact that he had let down the aquanaut team, himself, and me. I then tried to ferret out the reasoning behind his misbehavior. His answers were devious, specious, and loaded with rationalizations and self-pity. Finally, I asked him to look within himself and stand ready to talk with Scott and me tomorrow, anticipating dismissal from the program at worst, and retention as a surface support diver at best.

As a sad clincher, today I received a personal note from A's wife, noting that A would be on the bottom during his birthday and asking me to send him a cake and sing "Happy Birthday" over the communications system. How to answer a proud wife who has agreed to a most hazardous venture for her mate? Can I send this man back, branded as a quitter, or should he be retained in some minor capacity, perhaps to damage the morale of our remaining aquanauts? The decision is mine, and the isolation of command grows by the hour.

26 July 1965

Shortly after lunch, Scott, Walt, and I held a final conference regarding aquanaut A. When we agreed that it was best for the man and the program that he dissociate from the project, Walt came up with an excellent suggestion. Since a relief was required for diver Bob Sheats at the Torpedo Station before he could be released to us, let us send A up to relieve Sheats. The Torpedo Station would thus be getting an outstanding substitute diver, and Bob Sheats would be available to us early in the game.

And so it shall be. If all goes well, I will have A relieved in time to join us as surface support for the third run, when the situation and the personalities will have mellowed. He is a valuable diver who must learn to solve his problems without revolt against authority. When I gave my decision to him in private, he accepted it in good grace and with humility.

Today Paul Cohen of Sperry came by with news of his physiological telemetering system, which sounds very good indeed. We went over the prints, then checked out our various recording systems for compatibility. It now appears that we may be able to telemeter or record a free-swimmer's physiologic responses using three separate systems. Even if only one system works, we are ahead of the game.

Documentation, both before and after the fact, is vital to the success of any project of this magnitude. It is, however, a dreary pastime and no less frustrating in the light of certain knowledge that all carefully spelled-out procedures will invariably be subject to daily and often drastic change. Still, without a suitable frame of reference, chaos would be compounded. It is during this period of agonizing reappraisal that the long, painful moments of doubt put in appearance. Now we begin to consider not what we have carefully planned for, but rather what factors we have overlooked, perhaps because they seem so evident as to obviate need of documentation. We know that the risk of getting "bent" is the monkey on the back of every diver. The condition known as bends occurs when pressure forces into divers' blood streams part of the gas mixture that they breathe. When pressure drops inappropriately during ascent, the gas may emerge from the blood in the form of bubbles, causing such excruciating pain that the diver "bends" over in agony. The condition can be fatal. The responsibility for preventing any incidence of bends in our aquanauts during decompression lies squarely in my hands.

Two areas are of major concern to me at this point: the possibility of a communications breakdown and the probability of an aquanaut breakaway phenomenon. Against the former contingency, we have developed and are using the finest possible array of modern new equipment. Yet diver-to-diver and diver-to-surface communications have never been reliable, and it is difficult for me to believe that they will work this time, after so many failures in the past. Accordingly, I have calculated all my procedures to allow for total failure of these two communications links. The second hazard, aquanaut breakaway from topside control, cannot be fully predicted, but may be partially avoided by precept, prayer, and brainwashing my subjects. Further I cannot go.

27 July 1965

Scott and I had a poolside conference this afternoon to resolve the details of the day. He sees many similarities between the Man-in-the-Sea Program and the Manned Space Flight Program. The problems the engineers faced in designing the first closed systems for the Mercury spacecraft were very similar to the problems we face on Sealab. Both require the same types of machinery, in many cases the same chemicals, the same accuracy of monitoring equipment, and the same redundancy of backup systems. Scott sees a need for us to design and follow a well-planned selection and training process for the Navy's Man-in-the-Sea Program, as stringent as the program designed for space flight crews. The reasons for this are glaringly obvious. Considering the weight of seawater, the pressure differential imposed on the aquanaut diving to only 33 feet in depth will exceed that imposed on the astronaut walking on the surface of the moon. Again, comparing decompression problems, the astronaut can be guaranteed protection against bends through proper denitrogenation in all but the most severe accidental circumstances—such as rupture of the spacesuit during extravehicular activity. But for the aquanaut, exposed to high ambient pressures of inert gases over varying periods of time, there is virtually no guaranteed protection against bends, and the cure for this dreaded disorder is not simply a return to atmospheric pressures of sea level, but the application of added pressure, often many times that of the original exposure. It is a stunning fact that in the year when an astronaut was able to emerge from his capsule into the near-vacuum of space more than 100 miles above the earth, the free-diving aquanaut could descend only about 300 feet below the surface of the ocean.

Although Scott calls himself a "dry-land Colorado boy," he has been fascinated by the sea since his first tour in the Navy which, like me, he spent in Hawaii. "I began to do some diving," he said, "and it grew on me, as it does on some." The problems originating from isolation and confinement, experienced by both astronauts and aquanauts, don't concern him as much as the problems in maintaining environmental control of a closed ecological system. The astronaut has a relatively simple single-gas oxygen supply, whereas the aquanaut requires a multiple-gas mixture, synthesized in proportion to pressure and often having extremely low percentages of oxygen. Although these differences exist, control of the breathing mixture remains a common denominator of both programs. Scott noted that NASA is now able to simulate an entire range of environmental systems, as far as the atmosphere is concerned, and the Manned Spacecraft Center in Houston has a stable of simulators that our aquanauts would find incredible. NASA can accurately duplicate the conditions that the humans on the moon will find and can train their subjects in these simulated conditions. In fact, Scott said, by the time they get a crew of astronauts on the moon, some will already have been there for all practical purposes. But this degree of simulation, unfortunately, does not yet apply to our assault on the ocean floor.

Another concern we both share is the human–machine interface, where good basic design and fabrication result in a reliable machine that is safe to operate, easy to control, and reasonably comfortable to live with for long stretches of time, in confined areas. "Nuclear boats are great," Scott agreed, "although I think nothing today approaches the compatibility of the man and the machine that is found in modern jet aircraft and modern spacecraft."

In further extension of the comparison, we readily see that small-group interaction, captive atmosphere contaminant control, sensory deprivation or alteration, stress phenomena, and a host of other problems not directly related to the engineering obstacles harass both programs. If, right now, we had the capability to put free-ranging divers to a depth of 600 feet, we would open up the continental shelves of the world, which are 2 million times closer than the surface of the moon. We know great abundance of food lies there. But the implication of this fact has not yet been fully understood by the governments of the world.

1 August 1965

Westinghouse engineers Alan Krasberg and Jerry O'Neil have arrived with the Krasberg and Arawaks units, comprising a diver support system that is installed within the habitat. The system supplies a revitalized breathing mixture to the diver on an extended mission outside the habitat. The hand-carried equipment, which the engineer–inventors named Arawak after the Indian people of the Antilles, is in fact overengineered for our purposes.

The personnel transfer capsule is due here for refurbishing and final test procedures. Next, the deck decompression chamber will be delivered for final installation and checkout of the critical mating procedure. We await arrival of the two gas-jamming pumps and their checkout, then we're out of hardware and down to final details.

A call from Bob Worth came through to remind me of the TV show scheduled for nighttime taping. I was ready, if not eager, to comply. After all, it would only be 30 minutes of talk, scarcely a warmup for a dedicated raconteur like me. Then the other boot dropped. Scott Carpenter could not join me on the program. In a flash, my astronaut shield of protection vanished and I felt naked. After making a 220-foot dive on the site, Scott had to stay near a recompression chamber for the customary 12-hour period. That was as it should be, although I thought the dive was scheduled for a later time. Nevertheless, requirements of the Sealab program outweighed my personal considerations and I accepted the situation with ill grace.

There remained the problem of getting to the NBC studios somewhere north of Hollywood. Travel on the California freeway system is, to me, equivalent to plotting a trip to Mars. Knowing that I had a map in the glove compartment, however, I set out in good faith. Very quickly, I found that I had not reckoned with the problems of freeway travel. When moving at 70 miles per hour in a six-lane highway, it is difficult, even dangerous, to unfold a highway chart and plot a true course with accuracy. Indeed, it is damned near impossible. Quickly dropping off at the nearest exit, I glided to a service station, only to be directed to another, more complicated freeway. In due time I found the studio, and then the fun began.

To complicate matters, I had a number of important phone calls to make, not the least of which was to my long-suffering spouse on the occasion of our twenty-seventh wedding anniversary. At the studio, I was assured that telephone service was ready at my command. Somehow the apparatus never seemed to be accessible and, as no opportunity was provided, no calls were made. Fortunately, in view of my silence Marjorie assumed that I was at sea.

During dinner we determined that the show would be 11 minutes too long and set about pruning the operation. The next crisis involved my choice of uniform. Although I was wearing wash khakis and boondockers, I had had the foresight to bring along a dazzling set of the tropical white longs considered acceptable by the military for film appearances. However, dazzling white is anathema to color TV. With minor overhaul of the camera lenses and lighting, it was tolerated. The show went on and my commentary became a reprehensible body of opinion, subject to later review and analysis by naval authority. In retrospect, the opinions I freely volunteered to the TV audience are now a matter of record and I must accept whatever repercussions occur.

4 August 1965

Back on board the *Berkone* the galley is functional, our communications shack is nearly operational, and laboratory spaces are nearly completed. However, a fairly serious problem in connection with our operation grows more ominous by the day: the environmental conditions prevalent at the selected site for Sealab II. Because of the season, unusual sediment transport, or the prolonged siege of Red Tide (or for reasons unknown), the visibility at the site has remained near zero. Worse yet, the silt accumulation, all black, threatens to further obscure what little visibility there may be. It seems entirely possible that visibility will range from 6 inches to 1 foot throughout the operation.

This possibility raises very grave problems. If these conditions continue, as they have for at least two months, we will be deprived of nearly 70 percent of our human performance data, for lack of visual records. Additionally, motivation for prolonged undersea work will undoubtedly plummet if the aquanauts are required to exit into a black curtain of silt, consistently failing to do their appointed jobs, and leaving no record save that of failure. I had selected this site partly to present our aquanauts with varying degrees of water clarity. I do not, however, want to offer them a set of hopeless conditions around the clock. In addition, underwater photographic records are critical to the Man-in-the-Sea Program, and national TV networks must be considered. If the silt compacts, and if visibility improves, the problem will be minimized. But this has not yet happened, and our deadline approaches.

There is another possible course of action even at this late date. Months ago, before a final decision was made on the location, a hard and favorable look was given to California's San Clemente Island area, where we have an active Navy underwater facility. In addition, we have a permanent and completely stable moor in water depths to more than 230 feet, in clear water with a hard bottom undisturbed by weather. Finally, San Clemente can provide shore-based power and water. We could shift to that location with no more than a week's loss of time and with reasonable assurance of an operation that would yield maximum data and a good photographic record to boot. Balance against this the possible alienation of some of our scientific community, loss of some of our daily TV communications with the world, and the obvious admission that our planning did not allow for the adverse conditions that might occur on the edge of Scripps Canyon. It is a tight and difficult decision, and not mine alone. But still, as it is my project and my crew, I will probably make the final decision and it must be soon. Last-minute decisions are difficult at best and always subject to vicious review. The present condition of our La Jolla site stands as patent evidence of poor planning on my part, despite many months of photo survey, core sampling, personal dives near the site, and long interrogation of the Scripps divers. I can protest an act of God, a mysterious sediment shift, or the Red Tide factor—but to no real avail. Better to have an escape route, and San Clemente fits the bill.

10 August 1965

An influx of visiting consultants has swept down on the Outrigger Inn bringing in their wake a retinue of lesser luminaries and bag carriers. Most welcome are Drs. Perry Gilbert and Karl Sem-Jacobsen. The former, of Cornell University, is world renowned for his research on sharks and other elasmobranchs; the latter hails from Norway, is a neurologist of world fame, and a man of voracious and protean interests. A huge, gangling, loquacious, friendly Scandinavian, Karl envelops everyone in range with his charm and enthusiasm. His involvement in our project is hopelessly out of proportion to the scope and price of his contract.

As Karl's arrival on Sunday was delayed, Perry Gilbert and I began a long conversation in my room at the motel, pausing only to migrate to the dining room for dinner. All through the meal and for hours afterward, we talked—mainly about sharks. It was a remarkable evening and before its end I knew that the aquanauts were in for an absorbing lecture delivered by an expert on this sensitive subject.

On Monday, Perry and Karl were introduced to the team, then whisked off for a tour of the staging vessel, *Berkone*, and an outside look at Sealab II, which was temporarily buttoned up for pressure testing. They were outspokenly pleased with our state of preparation. Around midday, our tour was interrupted when Denzil Pauli exploded from building No. 39, waving his hands to halt us in our tracks. He told me we had a diving accident on the island. I raced to the phone.

A few days before, I had decided to send an aquanaut diving team to San Clemente for the purpose of surveying and photographing the 220-foot site I had selected as an alternate to La Jolla. Walt Mazzone went with the team as diving medical representative since I had to remain in Long Beach, and on Monday Walt had elected to dive with the first team all using open-circuit, compressed air rigs. At about 205 feet on the bottom, Walt began to experience the unmistakable buildup of nitrogen narcosis. Simultaneously, his regulator seemed to be delivering an inadequate supply of air. Recognizing the seriousness of the situation, and without immediate help, he attempted to return to the descending line; but instead he swam into the blissful, deadly cloud of narcotic oblivion.

Bill Bunton and his diving partner missed Walt and

turned back to find him floating face down about 3 feet off the sea floor. Together they carried the unconscious man to the descending line and began to boost him up the 200-foot run to the surface, making sure that his mouthpiece was in place and that he was breathing properly. Progress was agonizingly slow, as it always is with an inanimate diver. Bunton, realizing the urgency of getting Walt up, jettisoned his own valuable camera, and pulled the CO_2 lanyard on Walt's vest, inflating his jacket.

The ascent up the line then went easier and soon they were on the surface, where Walt quickly regained consciousness, though he was incoherent and shocky. He was immediately transported to the pier, thence to the chamber, where a precautionary table I decompression treatment was started, and a phone call was put through to me.

As Lew Melson on the other end of the phone quickly sized up the situation for me on Monday, with assurances that Walt, at 100 feet in the chamber, felt and looked well, I hitched up my galluses for a fast run to the island. Entirely aside from the fact that I must go to observe and be with my friend and coworker, it was imperative that I see the effects of this mishap on the other team aquanauts.

The plane was held 5 minutes while I set a near-record through Long Beach traffic. Thirty-five minutes after takeoff, I was at the diving locker, ready to enter the chamber with Walt. Within a minute I knew that the aquanauts were concerned, but not shaken. The situation had been handled with speed and skill by real professionals. In the chamber, Walt was in very good shape, with no discernible aftereffects from his misadventure. I stayed inside for the brief duration of the decompression. During this period, Bill Bunton told me, through an oxygen mask, of the loss of his precious camera and his keen desire to recover it. I was in sympathy. Without the camera, which he had willingly jettisoned to assist a diving buddy, his personal project was finished. The camera might be found if search dives were made immediately; tomorrow might be too late, given the tidal sweeps and the near-neutral buoyancy of the camera.

I outlined the situation to Bob Sheats and to Bos'n McCafferty over the telephone in the chamber and requested their collective judgment. The answer was clear. They would prepare two Sealab divers to go on a search as soon as the chamber surfaced. In a 30-knot wind and with rapidly failing light, Bob Barth and Scott Carpenter dropped over 200 feet, located the vital camera and brought it back up. That is the sort of performance that makes me proud of this gang and completely certain of their steady competence. We packed gear and sped to the waiting plane.

On Monday evening I shared a late supper with Perry Gilbert, Karl Sem-Jacobsen, and Karl's charming daughter, Anna, who had majored in physical education and looked it. A tricky idea began to hatch in my furtive intellect. Our men had physical training every morning—why not a new instructor? Scott Carpenter took to the idea, and early this morning, with the aquanauts in ranks for calisthenics on the beach, Anna emerged from her Mustang and took charge. Forty minutes later, the Sealab team was pooped, Anna was radiant, and any tension created by the recent accident had evaporated.

Perry Gilbert then began 90 minutes of the most interesting illustrated talk I have been privileged to hear on the subject of sharks. The aquanauts followed him intently word for word. As the question period began, I perceived an urgent signal from Lew Melson. With deeply serious mien, Lew told me in hushed tones that he had important news for me. The Secretary of the Navy had approved awards of the Legion of Merit and the Secretary of the Navy Commendation to honor our Sealab I participants. As I had previously been awarded a Legion of Merit in connection with trials of submarine escape, I was to be awarded a Gold Star. I felt strangely warm all over and had difficulty in clearing these foggy bifocals I've started wearing. The large number of awards to be made, after nearly a full year of consideration, is unusual: six Legions of Merit and the rest Navy Commendations. They are richly deserved, although I believe that I personally deserve no more plaudits. Given the free hand to work that I now have, I only want to do the job for an outfit I love. Since the day I was sworn in at the Experimental Diving Unit, I have known that the Navy would be my life. I have received exceptional treatment and unusual privileges, which I can only repay with my best effort. In all truth, my award should go to Robert D. Workman, Captain, MC, USN, who has done more to advance the Man-in-the-Sea effort, deep submarine escape, deep brief dives, and general Navy diving capability than any other man alive.

Tonight, as I lingered on these thoughts, Lew Melson passed on other news. Six divers would dive on locations adjacent to our selected La Jolla site, since there were indications that a slush-free bottom might be found nearby. If need be, the moor could be shifted a bit. In any event, we would start at La Jolla, and damned near on time. A long and fruitful conference was called and the schedule rescheduled. The rest of this remarkable day was anticlimatic, though I recall with pleasure an extra-dry martini at lunch and a late dip in the pool this evening.

But I come to the end of this day misty-eyed, the result of reading my wife's letter, with enclosures. Marjorie's note says she will manage to endure the long, silent separation, and the enclosure tells me of George, Jr.'s, triumphs as a water skier and Judy's first tremulous steps in managing her own life, with backward glances at her parents as if to assure herself that we're still there. It is not a nice thing to realize that although I think of my family often, I worry more often about the project. Perhaps it will not be so much longer.

5

Sealab II
(August–September 1965)

In August, after six months of daily workouts with the complex equipment designed for Sealab II, our group of eighteen military and ten civilian scientific volunteers fell into shape. The final phase involved long hours of checkout in all the hardware systems of the completed habitat. Living accommodations, compared with those of our first underwater project, were downright plush, and provisions for autonomous existence were more advanced. Umbilical cables from shore and from the surface support ship, *Berkone*, provided communications and TV links and data transmission coaxials, plus lines for power, fresh water, and emergency gas supplies. On board *Berkone*, my elaborate communications network was set up in a trailer. From this center I could communicate, through five channels, with any station on *Berkone*, any ship at sea, or any point on the mainland. I could even patch a communication between inner and outer space with this remarkable network, which largely depended on the integrity of the Benthic Lab—a complex system of multiplexers, switching, and self-repairing electronic circuits.

The brainchild of Dr. Vic Anderson, the Benthic Lab was developed partially under the aegis of ONR and designed to the highest state of the art of American industry. We tied our chariot to this star and although the star glimmered, it did not work. Multiple failures occurred and endless adjustments were required before the Benthic Lab performed satisfactorily. This did not stop us, but it hurt like hell. We had

Visitors are welcomed aboard Sealab II. (U.S. Navy photo)

to switch to a second system, running a coaxial TV cable to Scripps pier and recircuiting much of our onboard communications to maintain full capability via *Berkone* to Sealab's umbilical cord. More hours of work when all hands were already worn down. More than two hundred failures were detected and remedied in the Benthic Lab.

20 August 1965

More complications hit. The power transformer, an item quite separate from the Benthic Lab but essential to the power supply from the shore, tumbled on a steep bottom and flooded. I can operate without shore-based power backup but, even worse, it is now certain that we are off a bit in the site planned for mooring. Throughout the day, one team of divers reported a downgrade that cannot be tolerated for the habitat's location. I am certain that they were about 100 feet west of the prospective site for mooring that was found by Ken Specht's divers, but there has been no communication between these two dive teams. Since I have no control over either team, I can only ask for a conference between the two.

Over many years I have learned that recording the location of objects or specific spots on the ocean floor is an exceedingly difficult task. Accordingly, when Al O'Neal reported early this week that Ken Specht had found the site for mooring, I asked how he had marked it for accurate return. He assured me that he had perfect transit cuts on the site, give or take 5 feet. My heart sank, as such a fix relates only to the overhead surface vessels, not at all to the location on the ocean bottom, which should have been indicated by placing acoustic pingers. At this point, I will send Bob Sheats to dive with one of Specht's men to relocate the prospective site, which I believe lies to the east of today's dives. Bob's evaluation will decide the final location for me.

It occurs to me that it was only a year ago that I was invited to Admiral Galantin's office and offered my present job in Special Projects. Lots of water under the bridge since that day, some headaches, and a fair amount of fun. Upon reporting for duty, I learned that my full budget for the fiscal year was $138,000, of which $40,000 was already committed. Since that day I have managed to spend nearly $1.75 million.

I think, however, we have done well in the six months since we got the green light. We have built a remarkable $1.5 million complex, trained twenty-eight people to perfect pitch, kept more than three hundred technicians busy, and endured nearly a thousand hours of interviews on radio, TV, and in front of live audiences. We have also coped with nearly a thousand setbacks in the form of fabrication failures and human errors, have written twenty-five full-length papers, and have traveled a collective distance of well over a million miles. With our aquanauts ready, the weather excellent, and the sea quiet, we are ready to go.

28 August 1965

On a bright afternoon today, we lowered Sealab II into position southeast of Scripps Canyon before the eyes, ears, and cameras of a large assortment of newspaper and TV reporters, who marveled at the skill of our riggers and the measured cadence of our profanity. For the latter, I offered the excuse that we were either carried away, or were convinced that various pieces of gear were about to be. Today, the image of kindly, lovable, pipe-puffing Papa Topside was replaced by the reality of a ragged, dirty, foul-mouthed old bully. By midafternoon, Sealab II came to rest on the bottom, in less than ideal attitude (bow up, port list) but dry, with ports intact and all systems working. We made a single attempt to reposition to a flatter stance but came near to a serious pressure differential, and I decided to go with what we had.

The habitat was then entered by the first of three ten-member teams, each destined for a fifteen-day stay. I have resumed recording the essential story of courage, skill, and dedication of these few aquanauts in my daily Chronicle.

29 August 1965

In Bat Cave, North Carolina, the Sabbath is a day of rest, worship, meditation, and relaxation. On board the *Berkone*, however, our philosophy is much more avant-garde. We work on Sundays, or at least we did today. Yet work is a relative thing, for ours is a labor of love.

I rose early, back to the old Sealab I schedule, and climbed to the trailer to relieve Walt, who reported a quiet night, then pitched off in search of gainful work to do as a substitute for sleep. I settled to a vigil before the TV monitor, listening to the helium mutterings of the aquanauts as they rolled out of their bunks by ones and twos and made their way to the coffee pot. To the unaided eye, they looked like any other group of U.S. sailors, but I guess I see them through the affectionate perspective of a father or, more likely, grandfather. I found myself counting them anxiously and knew that, day and night, I would count those sheep for a long time to come.

Later today, with all assembled below for a moment of contemplation, I read the Sealab prayer over the intercom and was personally moved, though I cannot speak for the others. I wanted to let them know I was standing by and praying for them. Work ground to a halt and the calm of the evening watch set in, the time for leisurely communication, rumination, and relaxation. Scott Carpenter started off with the observation that as Sealab sits with a port list of about 6 degrees and the bow is up about 8 degrees, the dinner plates slide off the table and the aquanauts have to eat standing up. The drains don't work and the plumbing has backed up. The aquanauts have started to refer to the habitat as "The Tiltin' Hilton." So goes the first Sunday with the undersea project.

30 August 1965

The aquanauts are beginning to become acutely aware of the water temperature. They make sorties from the habitat only when fully clad, even if they are going to spend only a brief time in the shark cage. The wet suits that they wore for entry into Sealab II have not reexpanded and are difficult to don, as was predicted. The suits sent down inside the habitat quickly reexpanded but, being ⅜ inch thick, require a lot of lead to hold them down. We have a long way to go in the direction of divers' thermal protection, although the CO_2-loaded and electrically heated wet suits have not yet been tested.

Today comes a horde of scientists and helpers, each with

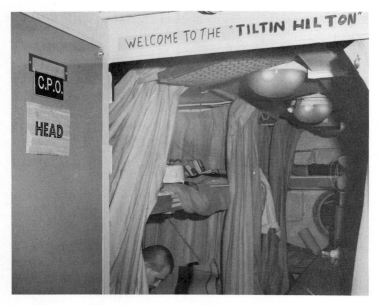

Because of the 6 degree upward angle in the Sealab II habitat, members of the first team of aquanauts dubbed their sleeping quarters the "Tiltin' Hilton." (U.S. Navy photo)

his own program, seeking time, space, and direct communication with the aquanauts. I am sure that all this is necessary, but it is not conducive to orderly progress on the bottom. Each interruption over the phones immobilizes at least two men, and slows them even more in their pursuit of useful tasks.

Also today, word comes again in regard to the Benthic Lab, the will-o'-the-wisp of the house of electronic marvels, the phantom of the Anderson lab, the miracle of multiplexing around which we built our own complex network of data recording. This same Benthic Lab will not arrive today, but perhaps tomorrow. However, the program is smoothing out, and infinitely faster than was the case after we got the men on the bottom during Sealab I. Our teamwork is vastly improved, morale both topside and below is very high, and performance excellent. In short, I am pleased.

Yesterday evening I called Marjorie, and was surprised that I remembered our phone number. Already, it has been nearly two months since we were together, with possibly two more to go. This is the last long one. From here on, I will sit topside as a visiting fireman. This job is wonderful and exciting, but it is time to get off the stage and into an orchestra seat—after October, of course.

Today, Scott and his team worked like dogs to complete cleaning their front yard, despite nearly continuous harassment from topside. Shortly I'll receive a SitRep, and I expect that almost all of the plan of the day will have been executed. Tomorrow we start the scientific program in earnest, and will, I hope, settle into a steady routine. I cannot help but reflect how secure and confident I feel, now that the aquanauts are on the bottom and the surface flail is over. I like to believe that this is a tribute to our long and careful years of work in the laboratory. The ocean is still a hostile environment, but we are taming some of it, bit by bit.

31 August 1965

A morning dive is scheduled as our first visit down to the habitat, ostensibly to swear in Billie Coffman for another hitch in the Navy. I have whiled away the time listening to our FM broadcast, intended to quiet the restless natives in Sealab II, and writing messages in pig-German to Dr. Sonnenburg, who had the watch. Despite his darkly knit brows, Bob has quite a sense of humor.

The late evening and early morning hours are the most pleasant, and often the most interesting, of all the time spent in the command shack. These are the hours when the hustle and bustle of work is over, and everyone is tired, satisfied with the fruit of their labor, and ready to chat about the big and little events of the day. These also are the hours when Walt and I are most refreshed and least irritable, since Walt sleeps by day and I by night, with early reliefs the rule. And so the pattern is set, a pattern that suits us both, and permits each to put in a 17-hour day.

In anticipation of the deep and fairly long dive to the habitat, I stoked away a sizable breakfast, then returned to the trailer for another three hours of intercom talk with Scott and other aquanauts, log-writing, and conversation with incoming scientists and visitors. Shortly it was time for our dive.

Under the watchful eye of movie, TV, and still cameras, Walt and I rolled off the end of the diving platform into the chill green of the Pacific, to rendezvous with the aquanauts 200 feet below. To spare our helium–oxygen gas supply, we rode the diving bell down a considerable distance before casting off to swim the final reaches to Sealab II. As we descended, dutifully popping our ears, the green water grew darker, then nearly opaque. Water temperature was about 49° F with very little marine life until we reached about 10 feet above the gray-black bottom, where we were surrounded by small fish of every variety. Ahead of us loomed the great white bulk of Sealab II, lying along an inclined sweep of ocean floor, her shark cage settled in the silt, the whole structure beginning to take on the patina of the surrounding seascape. I paused for a moment in my final descent, projecting an image of her appearance on the day of return to the surface, when she would look more like a scaly green denizen of the deep than the sleek white land creation we lowered last week. Hurrying on, I tunneled through silt behind Walt, removed my flippers in the shark cage, and struggled up the steps to the entrance well. Inside, unseen hands assisted me, removed my bottles, and I was inside Sealab II.

The handshakes and grins were as vigorous and warm as

if we had been apart for months, but each face wore a puzzled look as we began to talk. Down about four days now, the aquanauts had learned to accommodate their speech to the helium atmosphere, but Walt and I, freshly arrived, could not be readily understood. No matter; we were among friends. After a bit of a search, Billie Coffman's shipping-over contract was unearthed; the appropriate words were spoken and he was signed up for another hitch in the U.S. Navy in a ceremony rendered unusual by its exotic undersea setting. Minutes passed as we congratulated him, and then came the call from topside, "Time's up—get out!" The decompression penalty for disobedience of that command was too great, so we hastened to slip on the bottles and flippers while J. D. Skidmore, chief photographer, tried desperately to quell a balky strobe light that was ruining his still shots. Then we were out of the shark cage and swimming up to the bell for 73 minutes of decompression—a very long, cold hour and 13 minutes.

The crowning of this full day occurred just before dark, when the Benthic Lab arrived and was lowered beside Sealab II. Final hookup will take place in the morning. Following this, we start to build the ocean weather station. Since we are nearly through with house- and yard-keeping, we hope that, henceforth, more time can be allocated to scientific studies.

1 September 1965

I have just completed a series of tape recordings that I might entitle "The Sounds of Sealab II." There is a clearly defined tempo to the sound symphony arising from Sealab II, emanating from the intercom, helium unscrambler, FM radio, internal commercial TV, warning buzzer, Arawak pumps, and general background chatter. Add to this the topside cacophony of the twelve-station intercom, the pair of adjacent telephones, the relatively quiet Electrowriter, the commercial TV, bullhorn, and the ship's 34MC broadcast. Then include the babble of visitors within the trailer and the sound is complete. I have taped the entire symphony, from the still calm of the minutes of daybreak to the mad frenzy of the midmorning hour, through the ebb tide of afternoon traffic, and a return to evening's drowsy hum.

This afternoon, I am listening to the long two-channel tape I have made of the operation from day 1, to see if it is fit for reproduction and duplication. Thus far, I have identified only one spot that will not stand public exposure, an incident when I expressed my displeasure with the manufacturer of the recorder, which perversely reverses itself at the slightest provocation. My epithets would scorch a dragon's whiskers, so I shall delete this section for direct transmission to the manufacturer.

In all, this has been a trying day for the aquanauts. Far too many people got into the communications act, and the resultant breakdown in critical message relays shortly became manifest. There were failures to notify Sealab II of our impending trolley run, and failure to pass critical messages to Papa Topside. Long-winded conversations apropos of nothing useful often immobilized crew who had better things to do; unnecessary questions from topside required lengthy and meaningless answers. I sought to intervene to reduce the flow of babble, with notable lack of success, for by now the scientists' helpers entered the act. Having arrived on scene with untried and uncalibrated equipment only a couple of days before, they spent a good bit of time trying to bring their sophisticated gear to peak performance in a single afternoon, without consideration of any of the other elements of the program. For hours, these boy electronic geniuses held forth on every available channel to the habitat, the Electrowriter not excepted. Malfunction and delay followed trial-and-error as day follows night. All through this stood Dr. Sonnenburg gusseted with electrodes, rectal probes, and wires leading nowhere, resembling Prometheus Bound. At length I rebelled and cleared the stage. Dr. Sonnenburg proceeded solo with his cold exposure and recorded his locally acquired data, and the whiz kids were sent home to brush up on their circuitry. I blocked all access to communications and let the aquanauts have a well-deserved respite.

3 September 1965

This day started right and darned near went right all the way. Walt and I, on our joint early watch, talked for an hour and a half of the night's work and messages and of the personalities of this team. We are both justly proud of this bunch. There has been no slackening of motivation or enthusiasm for plain, dangerous, hard work. On the contrary, morale and teamwork have increased, if that is possible. How I am so fortunate as to have such an overall outstanding group, I cannot say. I am only thankful that it is so.

By daybreak, several of the aquanauts began to rouse, and a lively exchange of messages took place via the blessed Electrowriter. Some wished to know weather conditions topside, after noting that the weather on the bottom did not change much from day to day. Others shaped up some rather extravagant grocery lists and a miscellany of desired items, many of which had to be expunged from the list as not complying with our puritanical code of aquanaut behavior, coupled with our space limitations in Sealab II and a firm stand on hot-bunking. Nonetheless, the trolley line began what was to be a long day's work.

At this point, an issue that has become a matter of major concern to Walt and me reared its head. It is the matter of gradual but clear-cut invasion of the privacy of the aquanauts and topside control. It was my firm original understanding that the role of the psychologist teams at the Benthic monitor stations would be that of passive observers of activities inside and outside Sealab II. It is now apparent that outside long-distance phone calls are being monitored, which I will not tolerate. All outgoing calls made by the aquanauts are absolutely their personal affair. Further, Electrowriter communications between the aquanauts and topside watch-standers (Walt and me) are personal and private. I intend to call a conference today to clarify this point. Any violations will result in cutting off all Benthic communication lines, whereby all scientific disciplines will suffer, and fur will fly.

Today, the Scripps scientists proposed to place another weather station on the bottom. This time a unilateral decision was made to have the placement managed by the aquanauts. The plan was to inflate four bags attached to uprights, reach a nearly neutral buoyancy, then manhandle the contraption into place. The bottle supplied for this inflation was a single 38-cubic-foot container. When Tom Blockwick and I saw this setup, we exchanged furtive grins, since it was certain that this would not suffice to inflate the bags at 200 feet, no matter how well it worked at 20 feet. Doug Inman was so sure of himself, however, that we said not a word. Sure enough, later in the day an urgent cry came up from below asking for an additional source of buoyancy. With straight faces, we sent down a set of double 90s so that the show could go on.

Then came our first minor equipment casualty. The MK VI diving gear, which had a great batting average at Bermuda, had functioned flawlessly for seven days of rough usage with no failures. This afternoon, Earl Murray's gear flooded out, forcing a return to the habitat. The cause is as yet undiagnosed but, in a sense, I am glad that it happened. Earl was close to Sealab and not in danger, but tonight every aquanaut will mentally rehearse his casualty bill, knowing full well that such a problem could happen at a greater distance, requiring practiced skill to return alive.

The hostility of the black and cold ocean bottom off La Jolla cannot be overemphasized. Three modes of sudden death are always present when an aquanaut leaves the shelter of the habitat. Foremost is the possibility of becoming lost under conditions of near-zero visibility, always a consideration. The only haven for an aquanaut is a 48-inch hole in the bottom of Sealab II; to go anywhere else would be fatal. Another ever-present hazard is accidental buoyancy that would carry the aquanaut to the ocean's surface, meaning certain death. Finally, the dangerous marine life cannot be ignored; scorpion fish, which inflict horribly painful and incapacitating stings, literally blanket the ocean floor around the habitat.

The unanticipated problem of the anchovy-filled personnel transfer capsule remains unresolved. Scott does not want the lights turned out, but where there are lights, there too you will find anchovies, first by the gross, then by the barrel, and finally by the ton. It further follows that within hours you will find dead anchovies, putrid anchovies—reeking, distasteful anchovies. This bulky mess, having migrated up through the lower grating of the PTC, presents a real problem, since the PTC should at all times be ready for occupancy, which it surely is not at present. We plan to flood the chamber by a few feet, then attempt to blow the whole rotten package through the grating and skirt, and out to sea. After repeated sousing, maybe Scott will go along with shutting and lightly dogging the PTC lower hatch—the only possible solution. The cleanup is a diving job for which no one is volunteering. . . . As it now turns out a few hours later, Dr. Bob Sonnenburg was nominated for the PTC cleanup job, which he undertook with vigor and zeal, if not relish. About 90 minutes later, Bob and his buddy diver had done their bit for the cause, and the final cleanup team sallied forth to gather the last unwholesome morsels. On return, Bob wryly opined that the odor in the PTC was no more than an order of magnitude raunchier than in his berthing area, a remark that cast grave doubts on the house-keeping skills of the aquanauts. These doubts were quickly dispelled, however, when Bob loyally amended his statement to "a little worse than a pigpen."

Much more took place this day, but these few highlights may serve the purpose of revealing a typical daylight scenario on the bottom of the Pacific. Tomorrow is another Sabbath, but I must anticipate a bit of impiety here and there, conditions being what they are. Lest we become unctuous, let us remember that aquanauts are God's creatures also.

5 September 1965

A second Sunday for our aquanauts on the bottom, but somewhat different from a week ago. To me this seems to recall the opening chapters of Jules Verne's *Mysterious Island*, in which the castaways were so desperately concerned about bare survival that Sunday worship was confined to a hastily muttered prayer, and no day of rest was allowed. As they improved their situation, however, the inhabitants of the island gradually were able to set aside a day of rest and worship, to be maintained throughout their stay on the island. It is the same with our first team of aquanauts. Last Sunday, the urgency of critical house preparation did not permit the luxury of a Sunday holiday routine. Today, however, although the job is far from done, they all recognized the necessity of physical and mental rest. Strict communications silence was maintained through most of the day, save for about a half-hour reserved for church services.

The morning service, held at 1100, would have gladdened the heart and hackles of any preacher. Two hundred feet below, a completely captive audience of aquanauts huddled before the loudspeaker, under the all-seeing eye of a TV camera, which dutifully recorded every twitch, yawn, and shuttered eye. Above, I commandeered the public address system, stopped all ship's work, and had control for nearly a half-hour. From the opening rubrics, I could sense that we were together, giving thanks for the success of a dangerous but worthy cause. All the rest of this day, we dealt with one another a little less harshly; and as we worked, we rested.

6 September 1965

As if a day of rest was all that had been needed, the aquanauts hit this day with verve and precision not seen for a spell. Good humor returned; they were courteous and compliant. We swapped jokes on the Electrowriter and over the phones. The aquanauts even feigned to forgive me for (they said) commencing the venerable Twenty-third Psalm with the words "Papa Topside is my shepherd. . . . " This, of course, was a base canard, but I could not find it in me to be put out with them.

After a long SitRep and plan-of-the-day discussion with Scott, the morning series of sorties began. Work included outfitting the underwater weather station, continuing the fish census (too many scorpion fish for my liking!), and exploratory expeditions. Today's visitors included a reporter from *Time* magazine; a TV commentator; "Pete" Kidd, of the Royal Canadian Navy; Capt. Jack Kinsey; and many others. Team after team went out on time. The trolley line functioned smoothly. Skidmore took hundreds of feet of documentary movies, and work inside the habitat moved at a normal pace. Even the communications systems performed without failure. We visited at length as we planned the event of the afternoon—Dr. Bob Sonnenburg's birthday party.

For this event, Bob's wife, Pat, and his mother came aboard, after a long and somewhat bumpy ride from Quivira Basin. Pat, ebullient as ever, bore a large devil's food cake, complete with Sealab II decorations, all nested in a gay hat box. But she was near to tears when the box was opened, revealing a shattered masterpiece, made so by the incessant

Papa Topside, in a reflective moment, loads his pipe—his favored defensive maneuver. (U.S. Navy photo)

pounding of the *AVR-10*. Nevertheless, she bravely packed the cake and the cards in our pressure pot, posed for the TV and other cameras, and sent it on the long journey to Sealab II.

In the shark cage, Skidmore waited patiently to film the immortal scene of the festive pot sliding down the trolley, being disengaged, and being raised triumphantly into the habitat. Then, scrambling up the ladder and doffing his Arawak gear, he resumed his role of the day, covering the balance of the procedure with still shots. Bob Sonnenburg posed with the shattered cake, installed his twenty-eight candles, and unsuccessfully tried to light them with matches in the oxygen-poor atmosphere. Over the intercom, we sang "Happy Birthday," mercifully drowning out the remarks of his fellow aquanauts, and the party concluded with a two-way conversation between Bob and his lady visitors.

The last water event of the waning day was a submerged test of the electrocardiograph (EKG) monitor–telemeter system, using Fred Johler as a subject. In keeping with the success of the day, the damned thing finally worked, at a distance of 100 feet! After nearly ten days of repeated failures, that was indeed a welcome surprise.

In closing today's Chronicle, I feel that a word is in order relative to new developments in the area of underwater telemetering of biophysical data. Far too many data acquisition systems have been injected into the Sealab II program with salesmen's assurances that they are tested, will work the first time, are automatic and sailor-proof. In ninety-nine cases out of one hundred, this is pure, unadulterated hogwash. Almost without exception, the systems provided us by university groups and industry alike have been laboratory prototypes that have a reliability factor near zero even under ideal conditions of the cloistered campus atmosphere; they are years away from trustworthy operational application. Once on scene, the instruments require constant maintenance, calibration, and rebuilding by the host of technicians who accompany the brainchild, all of which ties up our critical communications, wastes our time, and clutters up valuable space in our cramped quarters. In addition, the instruments topside become seasick, as do the operators, and the sensing packages below don't like helium, or can't function in salt-water. In the end, in an effort to correct these major flaws, the technicians invariably proceed to add other unreliable and highly sophisticated components, rather than to strip the basic gear to its jockstrap for simpler operation. The result is chaos, short tempers, and interference with useful undersea operations.

If I have railed against the callous disregard of quality control on the part of American industry, I must also level an accusing finger at the university groups who seek to get their brainchild aboard a successful operational experiment, so they may rush into professional print with a first, gaily riding the coattails of more solid work. A plague on such philosophy, say I.

8 September 1965

Channel fever has set in for the aquanauts. Talk of the beach and upcoming liberty is rife; but little do they know how sparse their moments of leisure will be. Two days of hospital and laboratory investigation, one day and night of liberty, then back on watch as surface support divers. Come to think of it, the old principal investigator and his cohorts Walt and Tom, not to mention our master diver, are doomed to longer days of hormonal desuetude than any of the aquanauts. In fact, we will all have qualified for novitiate status in any of the many monasteries hereabouts, although we have received no invitation to join such a celibate group as yet.

Today, a great deal of energy will be expended on the man-killing task of rigging a preventer-anchor hookup to Sealab, presumably to prevent the habitat from falling into the Scripps Canyon. This possibility is so remote as to be unthinkable; nevertheless, the hookup is deemed desirable by personnel on the bottom, so it shall be. To date, however, we have seen the messenger line fouled on every damned object on the bottom, often to the peril of the aquanauts, and I am unhappy about the unilateral decision to install the requested hookup unnecessarily.

In addition to this somewhat useless task, a great deal of important work is being accomplished by the aquanauts. The ocean weather station is 70 percent complete, and all elements are so far functional. Marine biological experiments are going at a good rate. Excursion dives to the safety perimeter of 233 feet are being made routinely. The motor skills and strength tests are well underway, and other oceanographic studies are proceeding apace.

To a large degree, the first team have been cheated of the fruits of their toil, inasmuch as they have been stuck with the burden of the housekeeping duties as well as the problems of bottom equipment setup. The joys of leisurely sightseeing on the ocean bottom are not for them; even the simple pleasure of relaxation in a well-ordered undersea house is forbidden, for every waking moment must be spent in rearranging, patching, and planning another skirmish in an uphill fight. At long last, when the habitat, outside grounds and avenues, power plant, and outhouses are all in apple-pie order, Papa Topside will order them all, save one, into the PTC for a long ride back to the world of dry footing and bright sun, with scarcely time for a backward look.

Now comes team No. 2, hell-bent to produce a maximum of physiological and other scientific data, and seemingly oblivious to the preceding hard labor that provided their comfortable platform. This will be a period to test the iron self-control of Scott Carpenter, as he endures for a while the callous indifference with which these interlopers accept their new quarters and well-marked landscape. And if Scott should backslide a bit and permit himself the luxury of a group chewing-out or a mild temper tantrum, it will be accepted philosophically by those of us who are privileged to watch our colony grow from a floating point of vantage. In time, team 2 will commence their own uphill fight against the hostile environment and the strictures of Sealab II life, and the widening spiral will continue until the day of departure late this month.

Howard Buckner, chief steelworker, enters Sealab II for physiological tests prior to occupation by team 2. (U.S. Navy photo)

Finally, the third team will enter the now-venerable home of the aquanauts. I think that as they approach Sealab II near the end of their descent, they will first be struck by the appearance of their new house. No longer a pristine white with gleaming metalwork, Sealab II by then will have become a natural part of the sea-floor topography, a slate-gray mass surrounded and partially inhabited by swarms of marine life, blending so perfectly into the bottom seascape as to almost escape identification. Once inside, team 3 will note that the house has a definite lived-in countenance and gives evidence of persistent molding in the hands of her former tenants. Just as each generation of ancient cave dwellers left indelible and revealing marks and patterns of occupancy and culture, so will it be with Sealab II, the shelter, haven, and undersea cave of humans seeking dominion over the sea.

But shortly this impression will fade for all save Bob Sonnenburg, who will feel persistent nostalgia as team 3 enter to tackle the hardware and salvage problems for which they have trained so long. Under the leadership of Master Diver Robert Sheats, they will bring to the scene a degree of competence, efficiency, and effort that may seem to eclipse the capabilities of the previous groups, if only because of increased outside working times and proficient use of the tools of the salvage trade. Team 3 will be the salvors, the practical mule-haulers who carry on their work against any odds the sea may choose to set. They, too, will accomplish their assigned tasks and more. On that day in October when they file out of Sealab II one at a time, to greet and be greeted by surface dwellers, I will bet that each will take his own private farewell look and say, "By God, that was a good job, well done!"

6

Sealab II
(September 1965)

11 September 1965

This day, like most, started quite early. Often as not at this hour, the bath will be occupied by a rigger fresh in from a night on the beach, seeking to revive the flagging spirit with a cold rinse. But today my luck held, and it was not necessary to evict anyone from the washbasin or shower. I reveled in the solitary splendor of the head and its accessories and then joined Walt in the trailer.

The first call, from Washington, D.C., had to do with a story published today in the *New York Times*, which caused a bit of consternation in the high circles of the Navy. Last week I was asked if I would submit to an interview by a *Times* reporter. I made a special effort to comply with this request, since I have always held the *Times* science reportage in high esteem. The *AVR-10* came alongside about mid-morning last Thursday to discharge her daily first load of tourists, workers, and journalists, among whom was a bespectacled gent carrying a reporter's credentials from the *New York Times*. When he announced he was here to write only the straight scientific facts of the project, without sensationalism or reference to birthday cakes, I took him at face value. For nearly 90 minutes I answered his sparse questions with detailed and frank appraisals of the scientific views and achievements of our program.

As I spoke with the man, it gradually dawned on me that a certain substantive background was lacking. His questions were inane and naive. He had never heard of Sealab I or any other undersea venture and did not know the continental shelf from the third aisle in a Safeway supermarket. In growing horror, I realized that my scientific discourse was beyond the comprehension of the reporter facing me and suggested that I could blue-pencil his copy to assure accuracy in the article. However, he brushed off my offer, saying he would telephone me if he encountered anything he did not understand. As he left the ship, I yelled after him that my greatest fear was for the stuff he *thought* he understood. Too late; he departed without a backward wave. I thought of a good remedy, but *Berkone* has no small-arms locker. The egg was on its way straight for the fan.

According to the article that subsequently appeared, our program had nearly been terminated by failure of thermal protective garments; the aquanauts, already debilitated by painful and incapacitating ear infections, were developing a new and unexplained blood disorder and lived in constant fear of rolling into the depths of Scripps Canyon. I was directly quoted as saying that no useful information had been derived from last year's Sealab experiment. All of these errors had crossed the desk of a *New York Times* science editor without question or any check of validity. An official protest was lodged at my urgent request, but no newspaper will repudiate an entire article, and the damage had been done.

Perhaps I should not deal too harshly with the reporter, but rather with the editorial judgment that permitted his assignment to a scientific project. Today, the article was posted on our events blackboard for all visiting journalists to see and digest with care, along with my own earthy comments.

Too much time has been spent in discussion of the frailties, principles, and proper functioning of the pneumofathometer system designed for Sealab II. A pneumofathometer is a stock-in-trade item with all hard-hat divers. Its principle derives from the observations of Sir Robert A. Boyle: If a rubber hose is filled with air and extended to any water depth, a gauge reading the pressure of the air in that hose will tell exactly how far down the hose the air column extends. If the air pressure in the rubber hose is gradually increased by admitting more air, the air column in the hose will extend finally to the end of the tube. It is that simple. It became complex, however, as Walt and I alternately tried to explain the system to Scott Carpenter. As Scott first put the problem to us, he had tried the pneumofathometer and it had indicated a depth 20 feet shallower than the depth gauge. Of the two, Scott said he considered the wrist gauge accurate and the pneumo wrong.

I replied that the pneumo, properly used, is always as accurate as its own topside gauge, whereas wrist gauges are not consistently trustworthy. At this point, Scott sensed a criticism of his manipulation of the pneumo, which was frankly the case, and launched into an amazingly detailed description of each nut, bolt, and air molecule in the thing. Next came requests for additional gauges, reducers, clamps, and so forth. I tried to assure Scott that he already had all the essential elements for proper operation and needed no added paraphernalia. Thus encouraged, Scott recapitulated the operating procedures, this time including a fact that had previously gone unnoticed. It seemed that Scott had been flowing gas at a good rate through the pneumo tube, and taking readings *while the gas continued to flow*. Such a reading is of no value, but I could not make this point clear to Scott, who talked about "line loss," which means nothing in a static, or stationary, column of gas. As time flew by, I strongly recommended turning the whole job over to the master divers topside, whose cumulative experience with pneumofathometers totaled forty years. This suggestion was unacceptable to Scott, who countered with further requests for other hardware. I turned the phones over to Walt, who was still talking and listening when I quit the scene. Later, the subject was simply dropped for the time being.

This is not to say that Scott was being rigid or obstinate,

although these adjectives did present themselves to my battered psyche. Coming from aviation training, he believes in instruments with dials, while we underwater buffs have learned that divers' wrist gauges are not to be trusted, but a pneumofathometer, properly backed up by a recently calibrated gauge, is foolproof. Even in this uncertain world, some things must be taken on faith and, until Boyle's Law is repealed, I shall insist on pneumo readings as my criteria of depth determination.

As much as anyone, I favor a full explanation of all procedures and decisions, but when such discussions reach filibuster proportions, the principal investigator must say what shall be so; and so it shall be. Tomorrow, the Sabbath, I intend to preach on the subject of faith.

During the rest of the day, we had a few minor happenings. The personnel transfer capsule—the single emergency escape chamber available to the men below—was scheduled to be raised and placed on deck for a thorough cleansing and deodorizing, as the anchovy remnants had not all been removed and the stench was unbearable. I remarked to Bob Sheats that I could never recall making a drastic and unwilling decision such as this without later saying, "Thank God I did!" Invariably, upon close inspection when the deed is accomplished, one will find three good reasons for having done so. As we prepared reluctantly to raise the PTC, we learned from an inspecting diver that a shackle pin was missing from one leg of the lifting bridle. That could have been a serious matter during a fully manned lift-off. As we repaired the damage, I said to Sheats, "That's No. 1."

The lift then went smoothly and the PTC was deposited in the retaining rack without mishap. Bob Sheats walked slowly around the hull, then called me over. "And there are Nos. 2 and 3," he said. Sure enough, one fitting on our emergency escape system had been snapped off by our free-swinging bridle on the bottom, and the underwater phone connection had been battered beyond function. It took all night to repair the damage and deodorize the PTC, but the job was completed this forenoon and she was returned to the ocean floor in good working condition.

The balance of the day was spent in transfer of groceries, equipment, and personal belongings. Tomorrow, we exchange teams 1 and 2, the former to endure the hazards of linear decompression, while the latter group enters the hostile, dark, cold environment.

Walt and I ask for God's special help in the hours and days to come.

12 September 1965

In terms of all-around accomplishment, it would be hard to beat this day of work. At first light, I was pleased to note a flat, calm sea, an unusual east wind, and a fairly clear sky. This would be the most dangerous day, the early part of the decompression experiment. Walt and I rejoiced together over the promise of ideal weather and agreed on details of the decompression procedure. Breakfast came while Walt went below for an hour's sleep. I readied three divers of team 2 to go below for 6 hours of intensive physiological studies and to keep Scott Carpenter company while awaiting the transfer of teams 1 and 2.

Divers Bob Barth and Howard Buckner entered the water at 0845, with our blessings and farewells. Soon after, we began a long series of final checkouts of the PTC, location of the fish cages, and bottom sorties, interspersed with many more trolley runs to transfer personal effects, dirty clothes, and a number of obscure items that, unclaimed by Sealab II and topside alike, seem destined to ride the trolley pots forever.

The *AVR-10* arrived alongside and deposited a gaggle of journalists, military persons, TV camera crews, and one very seasick female passenger looking green and wretched. Now the deck bustled with activity. In a secluded corner outside my trailer window, Dr. Sem-Jacobsen's three ravishing helpers chattered in Norwegian as they fussed over five beaming aquanauts, placing electrodes in appropriate areas of scalp and chest, while ignoring requests for Swedish massage and the like. On the fantail, groups of reporters queried anyone who even looked like a diver, while my trailer was overrun by a motley group of the curious.

By noon, all was ready for a final inspection of the PTC prior to loading team 1. Scott Carpenter, who had scheduled himself for this job, dropped through the hatch to make the short sortie. Less than a minute later, he appeared on the TV screen, being assisted to the berthing area by two fellow aquanauts. The bad news came soon enough. Scott, in passing through the shark cage, had inadvertently touched a large scorpion fish with his left hand. The reaction was swift and venomous. Three or four spines pierced Scott's index finger, resulting in excruciating pain, followed by rapid swelling and disability. As shock and respiratory failure were quite possible, Dr. Sonnenberg went into the plan of action we had previously laid out with care. Scott was placed in a head-down position on his bunk, intravenous antihistamine was injected, a decent dose of pain-killing narcotic followed, and cortisone was given intramuscularly. Meanwhile, the hand was chilled and antishock medications readied for instant use.

Over the public address system, I issued a bulletin explaining the cause of the holdup and the nature of Scott's injury, and then offered glib phrases of reassurance that I did not personally feel at the moment. One anxious hour passed, then another, and Scott was out of danger. We decided to leave him as team leader on the bottom as scheduled. Bob Barth was selected as acting deputy team leader, and the show resumed.

In a smooth 40-minute operation, nine of our team 1 aquanauts were lifted from the ocean floor in the PTC and transferred to the deck decompression chamber on board *Berkone* to remain under continuous decompression until surface pressure was reached.

Two hours later, delayed by an unforgivable design error in the upper hatch of the DDC mating trunk, which had to be held shut by means of an extended automobile bumper jack, the PTC had again descended to rest outside the shark cage and a new set of aquanauts began the descent to Sealab II, one pair at a time. As I waved them a brief farewell, I marveled that the scene was by then a smooth routine and no longer a spectacular event. Tonight Walt and I will go with little sleep as we watch over our decompressing divers, but by midnight tomorrow, the even tenor of the operation will be resumed.

My church service earlier today contained a message relating to the power and comfort of faith. At this midnight hour, I believe my words all the more.

13 September 1965

This was a day of waiting. After three hours of catnap, I relieved Bob Thompson until noon, slept another hour, then prepared to lock into the decompression chamber with Dr. Karl Sem-Jacobsen to prepare for visiting team 1 and standing by while he took electroencephalograms (EEGs) on all aquanauts. The ebullient Norseman joined me in the outer lock, babbling questions as we descended, and incidentally forgetting to equalize his ears. Shortly, his pain was sufficiently severe to shut his mouth for a bit, but I kept flowing in the gas regardless, as I grimly contemplated his misery.

Prior to pressurization, I had sought to explain the basic principles of "ear-lifting," but Karl interrupted, "Not necessary, Georch, I have flown the X-15." Flying the X-15 is without doubt a worthwhile accomplishment, and shame to the one who dares gainsay it; but the maximum range of the

X-15, in terms of pressure differential, is less than that experienced in a dive 33 feet down, and we would pressurize to at least 60 feet. Sensing that logic would not prevail, I had grumpily ceased advising that he practice equalizing his ears and loaded Karl into the outer lock without further ado.

Now, at pressure equivalent to 30 feet, he was in trouble. His elongated Nordic countenance grew longer by the second; tears welled in his eyes; great veins corded in his red neck as he blew vigorously on his honker; and all to no avail. Remorselessly, the pressure continued, causing Karl's eardrums to bulge inward. He danced from one foot to another, as if barefoot on hot asphalt, tugging at his earlobes and emitting Scandinavian yelps that were drowned in the 112-decibel noise of incoming air. At length, Karl's white hair arose on his pate and trembled like the vibrissae of an eager insect, while his bristling eyebrows semaphored distress. In compassion, I secured the pressure and inquired after his health. Karl's response was a vigorous nod of the head, a slow foxtrot, and a word that sounded like "fine." In a moment the inner hatch opened, the jig ceased, and we entered the dank, sweltering, sewer pipe peopled with dripping aquanauts.

The next 70 minutes were spent taking EEGs, screaming Norwegian expletives to the technicians outside, praying in silence, and conversing with the aquanauts of team 1. When our brief visit ended, Karl and I decompressed at the 10-foot level for a period, then returned to sea level. I resumed the watch in the trailer and Karl returned to his cabin to nurse his ears and charge his recorder. Much later, he declared cryptically that the run had been a valuable experience. I could truthfully say as much for any close brush with disaster.

Later, after an hour-long SitRep with Scott, I was standing by the outer lock with Walt when midnight struck. The hatch opened and the disheveled team 1 aquanauts streamed forth to face the TV lights and press. Somewhat later, we agreed to meet on the O deck for a formal press conference in the morning. And so to bed, or so I thought.

Nearly midway in my siesta, I was awakened by Senior Master Diver Dan Price with news that three aquanauts had reported knee pain. In the diving locker, I was faced with three mildly apprehensive, recently decompressed aquanauts. Each complained of moderate, nonescalating pain in the muscles of the lower thigh, aggravated by motion and nonboring in character. It seemed significant to me that they were three of the four characters who had spent the last 5 hours in the deck decompression chamber on their hunker bones playing cribbage. In my drowsy state, and with a natural aversion to many more hours of treatment watch, I made an immediate diagnosis of genus *cribbaticus*, and ordered a deep sleep for all hands. I had guessed right; by daylight, they were cured of their affliction. Obviously, it never pays to make a hasty decision, however clear-cut it may seem, and especially if it threatens the rest and recreation cycle of the attending physician. I too returned to bed and slumber.

16 September 1965

There will be no need to simulate any casualties in Sealab II, at least for the time being. Even as I wrote yesterday's Chronicle, events were taking place that would test the mettle of a Spartan.

First, Scott Carpenter declared Glen Iley not to be in sufficiently good physical condition for MK VI diving. I concurred, as Glen had failed to live up to his promise of last winter to get in shape and lose weight. By the time I had seen him at Long Beach, it was too late to train a replacement, so I could only hope that calisthenics would help. Indeed, there was a serious question in my mind as to whether he could pass through the 48-inch-square entrance hatch to Sealab II. As an aquanaut at risk, Glen Iley will probably be designated for a sedentary existence in the habitat, with occasional forays to the shark cage or to retrieve the pots and baskets from the trolley. I cannot overlook the potential for serious injury to the man himself.

Next news was more serious. George Dowling, who had been virtually living in the experimental heated wet suit, appeared to have developed a sensitivity to it, and was experiencing swelling and rashes of the extremities, together with an apparent infection of one foot. Bed rest, antihistamines, and supportive care were prescribed.

Then came news of frightening importance. On a swim yesterday afternoon, one aquanaut, wearing insufficient weights, became uncontrollably buoyant and was saved from a fatal ascent only by quick action of his diving buddy, who hauled him down and back to Sealab II. I realize that our team members do not need to be so heavy that they must crawl among the thousands of scorpion fish on the bottom, but they must know that a state of positive buoyancy may result in a fatal rise to the surface.

Following this entry came a few more clear-cut indications of carelessness. One aquanaut returns to the entrance to ask if the control block of his diving gear is open! Still another diver loads his CO_2 absorbent canister with a *sealed* package of Baralyme absorbent material! My suspicions of gross laxity are plentifully confirmed. The system of buddy checks has begun to go by the board, and I recognize the first alarming signs of aquanaut breakaway phenomenon.

After discussion with Scott, I was assured of a rigid tightening of the rules. Still, it came as no surprise to me when Buckner received a sting on the foot from a scorpion fish or when a deep (250 feet) excursion dive was started on half-filled bottles, with no reports given of exit and entry times. Daily, there is increasing evidence that strikes terror in the heart of the topside watch-stander. The intercom calls that go unanswered for long minutes; the Electrowriter messages ignored for as much as a half-hour; and the situation report due at 2000 but not given until 2300 hours, often incomplete: all are warnings that I cannot ignore. At the risk of alienating Scott as team leader, I am forced to insist on compliance with my orders as principal investigator, without an interminable argument or even justification. The day has not yet arrived when the judgment and planning of topside control can be ignored or substantially modified by the subjects in the habitat.

This morning Tuffy, a Navy-trained dolphin, was brought out for a command performance before the press and TV observers. He lolled on or near the surface, bestirring himself only to snap up hunks of frozen mackerel and to leer at the audience, while his human playmates below shivered in cold water outside Sealab II, sounding their homing buzzers for him. Half an hour later, after an unscheduled performance near the fantail of the *Berkone* for the spectators, Tuffy flipped over and headed straight down for Sealab II. Four and one-half minutes later, he was on the surface, looking disillusioned and anxious to depart the scene, which he did. Shortly, the aquanauts reported that when Tuffy had reached bottom, he had taken a horrified look at the array of hardware on the ocean floor, then fled ignominiously. Tomorrow, we will try again. Tonight, however, I have an even higher esteem for the bottle-nosed dolphin; the Sealab II complex scares the hell out of me, too.

17 September 1965

It is apparent that our aquanauts need better protection against the scorpion fish that literally blanket the sea floor around Sealab II and are packed into every cranny around the habitat itself. Eradication or even thinning-out techniques are obviously impossible; we must therefore provide some foot protection for the diver, such as an inner sole to be fitted into the fins or attached to the suit soles. Tomorrow, we will check out the feasibility of putting protective soles or partial soles on the suits, which are not so equipped. This aspect of the hostile environment represents more of a painful nuisance than a real threat to the aquanauts, although a sting on unprotected flesh at some distance from the laboratory would be very serious. In any case, some ingenuity is required to reinforce the wet suits against these noxious pests.

This morning Tuffy lived up to his advance billing when he answered a buzzer call from an aquanaut 205 feet below. With an insouciant flip of his tail, Tuffy took off. Within 90 seconds, he greeted aquanaut John Reaves and carried a safety line from Reaves to Ken Conda, a "stricken" aquanaut some 60 feet away. Finding lost divers could be Tuffy's most important assignment. The dolphin really performed, making seven payload-carrying runs before we called it quits, suitably impressed.

Three days ago, I had the pleasure of a 30-minute underwater frolic with Tuffy. Our friendship was immediate and, I hope, enduring. While Bob Sheats shot photos, I scratched Tuffy's back and stomach, which he genuinely enjoys. As before, I was struck by the look in his eyes as he gazed at me, 40 feet down. It was the same look I have seen in the eyes of very old and wise people as they watch young children at play. Somehow, it is not right for the principal investigator of a million-dollar scientific project to be self-conscious in front of a ten-year-old *Tursiops truncatus*. I must also confess to sharing with others who work with Tuffy a conviction that, in his mind at least, it is he who is training us—and not finding it a simple job. In truth, Tuffy has taught us to feed him when he says certain chattering words or makes appropriate gestures. The fact that we enjoy his noises and actions is of no importance, as he seeks continually to improve the repertoire of those stupid men–fish who can't even talk at 3,800 cycles and who swim at only 0.8 knots.

By conservative estimate, we have had failures in nearly a

dozen TV cameras so far in this project, even though we are using the finest equipment available and the gear has constant and superb maintenance from the manufacturer and our own technicians. Being underwater cameras, they are in fact impervious to water and high pressure. They are not, however, designed to stand high pressures of helium, a gas that diffuses through almost any substance; consequently, the TV cameras leak inside Sealab II, become internally pressurized, and then fail. After battling the problem these many days, a flash of genius struck me. The TV cameras were designed to work underwater; therefore, let us put them in seawater outside Sealab, peeking in through the portholes. Dizzy with the brainstorm, I ordered an external camera carried to a port for demonstration. Within an hour the job was done, with complete success. Tomorrow, we'll have trouble-free TV, I hope.

Tonight the boys below are dog-tired and unquestionably irritable. Likewise topside. I prescribe a full night of sleep, and here I go.

19 September 1965

By agreement of subjects and topside control alike, this is another Sunday of rest, recreation, and repair. Consequently, no reveille was called and an air of indolence pervaded the otherwise busy TV screen. Topside, we maintained a relative communications silence, lolled in the sun beside the trailer, and caught up the log and other despised bookkeeping chores. The sea was calm; our solitary sea lion slept on a spud buoy by our bow, and the clear blue water had at last come in from the Pacific depths to improve visibility and raise our spirits.

Around midmorning the first boatload of visitors arrived, made up largely of old hands, Navy types, and beachcombers. We plowed through the amenities while I prepared for the morning church service, scheduled on the air and under the water at 1100 sharp. Search though we might, no church flag has come to light on board this old barge, nor do we have a bugler to blow church call. I must make do with our public announcement system, which has all the characteristics of a circus ringmaster's horn. All things considered, however, this may be an appropriate call to worship.

After rendition of a skeletonized version of morning prayer, I talked of the example set by my old and dear friend, the Reverend Richard Vause, who truly walked by faith and not by sight alone. His philosophy, "can do, can help, can stand anything," was an inspiration to me during many of the difficult days at Bat Cave. From the opening sentences of the call to worship, I was again in the Church of the Transfiguration, and the story of Richard Vause flowed readily before my mind's eye and out of my mouth. These memories make Sunday services a pleasure to me and, I hope, to my captive aquanauts. A little short on basic theology, perhaps, but the lesson is always there to be learned.

By noon, all preparations were complete for lowering the tripod TV camera outside Sealab II. A great deal of organic matter was in the water, and within minutes the scorpion fish began to congregate until the ocean floor was blanketed with the damned things. Shortly, a pair of black-suited aquanauts swam into view on the screen, removed the cable, which had inevitably become fouled on the tripod mount, and drifted off to an unnamed rendezvous in the murky depths beyond.

The fouled cable deserves an additional word. Certainly, if Murphy's Law (If anything can possibly go wrong, it will) holds true on the earth's surface, one must consider that in undersea situations, the law is true at least to the third power. Since beginning this project, I have faithfully recorded the fate of every object lowered to the bottom. Without exception, each and every item has become fouled, sometimes on lines at least 50 feet distant. The question is never "Is it fouled?" but rather "What is it fouled with?" Surely, the fouled anchor device that we proudly wear on our uniform caps speaks a world of truth for naval operations, past and present.

Another curious fact is that objects on the ocean bottom at a known location are not in fact there, but often are found some distance away and on a new azimuth. Although we have been moored over Sealab II, an object nearly the size of a small submarine, for nearly a month, we still don't know precisely where the habitat lies, nor on what heading. Surely, one day, we will build locating devices oriented only to ocean floor topography.

After lunchtime, the tempo of underwater life picked up, although still below the daily norm. Electrically heated wet suits were tested, TV cameras installed, marine biological studies continued, and psychometric arrays laid out. By the end of the day all aquanauts were again at rest, and the quiet watch was turned over to Walt, who so often knits up the raveled sleeve of care through the long night hours.

20 September 1965

Of all the hours of the working day, those in the early morning watch I share with Walt are unquestionably the best. Conversations with the aquanauts are limited to terse, often humorous, notes on the Electrowriter, and it is pleasant to observe the quiet scene on the TV monitor. Meanwhile, plenty of time to shoot the breeze with Walt, to see the program at a distance, and to swap ideas without interruption or kibitzing. These easy hours take me back to the predawn watches that Walt, Bob Workman, and I shared in 1958, as we toiled weekends over the precious rat experiment that led to our present operation. In those days we laughed at ourselves for being presumptuous, yet certainly we all felt that the present scene would come to pass one day.

Two items are of some concern to Walt and me. The first has to do with a minor dichotomy between topside control and the Sealab II team leader and crew; the second relates to minor physical complaints that emanate from the habitat.

Scott Carpenter was selected as training officer and team leader for the first two teams because of his excellent powers of observation, traits of leadership, positive attitude, and infinite curiosity. These fine qualities are all desired in our leaders, and there was never doubt of their presence in Scott. On the bottom, however, where autonomy is stressed and personal responsibility of the team leader is a heavy burden, some of his sterling qualities have become a source of irritation and occasional outright friction between topside control and the team leader. The problem is magnified with the passage of time and the accumulation of frustrations and personal discomforts that attend such a long stay in a hostile environment.

In any event, Scott's natural curiosity seems to generate excessive questions; he challenges every decision we hand down by demanding infinitely detailed explanation. In itself, the questioning is no great problem, but the lengthy defenses of each topside decision consume time that is precious. Scott even challenges some decisions relative to atmospheric mixtures that can be made only by topside control. This morning I devoted precious minutes to a completely technical explanation of why I ordered a change from 4.25 percent oxygen to 3.5 percent. Although I did not begrudge the explanation, it is a fact that some of us have devoted several decades to learning our business in high-pressure physiological programs, and our judgment must be taken on faith.

The second matter of concern to Walt and me is complaints that the atmosphere "just doesn't seem right," combined with reports of a high incidence of headaches. These vague but disquieting factors do not lend themselves to easy diagnosis or correction. We are using the most sophisticated techniques available to assure minimal toxic contamination of the breathing mixture, and our daily physiological program looks at nearly one hundred parameters of body function and states of health. As neither of these watchdog systems has uncovered the slightest cause for alarm, we are forced to assign a psychogenic factor to the complaints, while maintaining a steady vigilance. Nevertheless, it is disquieting.

Team 3, composed of divers with experience under more adverse conditions and led by a different personality type, may yield a markedly different portrait. Time will tell very soon.

23 September 1965

This morning started well enough, with a splendid situation report from Walt, marred only by the fact that oxygen control had got out of hand during the wee hours, permitting a rise above my experimental level of 3.5 percent, which I had hoped to maintain for 48 hours. Still, this was only a matter of grave concern to me, not a truly killing item. I puttered about the trailer, alternately cursing the dunderheads who couldn't watch their gauges and filling the pages of a fresh log book with minutiae. The work of the day continued, with the aquanauts making plans for another attempt at the 266-foot mark.

Now came a long discussion of the use of the pneumofathometer as the final criterion of depth attained. I strongly urged that one pneumo—our own, topside—be used. This met with countersuggestions that the private pneumo of Sealab II, long since discredited by us, must be taken along, presumably for sentimental reasons. Out of weakness, I capitulated. Almost as an afterthought, I was then told that communications with topside had been lost in the PTC, which, of course, serves as the only emergency escape vehicle. Further, I was advised that a few fish had infiltrated a loose area in the fish screen, and that the odor in the PTC was a bit high. As I framed an explosive reply, the aquanaut dropped the phone and took off on the deep sortie.

My Sunday sermon just might be taken from the *Book of Job*.

7

Sealab II
(September–October 1965)

Lewis B. Melson, Captain, USN, ONR, project director
H. A. O'Neal, ONR, program director

The Aquanauts of Sealab II
Robert A. Barth, chief quartermaster (DV), USN, team 2
Howard L. Buckner, chief steelworker (DV), USN, team 2
William J. Bunton, experimental test mechanic, Scripps
    Institution of Oceanography, team 3
Berry L. Cannon, electronics engineer, U.S. Navy Mine
    Defense Laboratory, team 1
M. Scott Carpenter, Commander, USN, team leader, teams 1
    and 2
Thomas A. Clarke, graduate student, Scripps Institution of
    Oceanography, team 3
Billie Coffman, torpedoman first class (SS) (DV), USN,
    team 1
Charles M. Coggeshall, chief gunner's mate (DV), USN,
    team 3
Kenneth J. Conda, torpedoman first class (SS) (DV), USN,
    team 2
George B. Dowling, research physicist, U.S. Navy Mine
    Defense Laboratory, team 3
Wilbur H. Eaton, gunner's mate first class (DV), USN, team 1
Arthur O. Flechsig, specialist oceanographer, Scripps
    Institution of Oceanography, team 2
Richard Grigg, graduate student, Scripps Institution of
    Oceanography, team 3

Glen L. Iley, chief hospital corpsman (DV), USN, team 2

Wallace T. Jenkins, equipment specialist, U.S. Navy Mine Defense Laboratory, team 2

Frederick J. Johler, chief engineman (DV), USN, team 1

John J. Lyons, engineman first class (DV), USN, team 3

William D. Meeks, boatswain's mate first class (DV), USN, team 3

Lavern R. Meiskey, chief shipfitter (DV), USN, team 3

Earl Murray, laboratory assistant, Scripps Institution of Oceanography, team 1

John F. Reaves, photographer first class (DV), USN, team 2

Robert C. Sheats, master chief torpedoman (DV), USN, team leader, team 3

Jay D. Skidmore, chief photographer (DV), USN, team 1

Robert E. Sonnenburg, lieutenant, MC, USNR, teams 1 and 3

William H. Tolbert, oceanographer, U.S. Navy Mine Defense Laboratory, team 2

Cyril J. Tuckfield, chief engineman (DV), USN, team 1

John M. Wells, research assistant, Scripps Institution of Oceanography, team 3

Paul A. Wells, chief mineman (DV), USN, team 3

25 September 1965

In the final hours of bottom stay for team 2, two aquanauts made a fast run to the canyon edge for a short period of wonderment before returning to the drab surface of the world. What a shame that a full twenty-nine days had to be spent in a black, cold mudhole before reaching an important goal of the program—the awesome Scripps Canyon—which could have been achieved days ago, save for the interference of time-consuming programs that held promise of neither success nor merit. Indeed, if I were to eliminate any individual elements of the Sealab II schedule, it would be those psychomotor studies that had little connection with the reality of underwater living. I think real insight in the psyche of an aquanaut is to be gained by listening to a description of his first view of the canyon; aquanauts cannot inspire such insight by filling out questionnaires or fiddling with gadgets that resemble children's toys and have an even shorter life span in the sea.

A description of a typical sortie to the canyon might go as follows: After the tedious and tiring process of struggling into unyielding wet suits inside Sealab II, my aquanaut swim-buddy and I drop into the black coolness of the exit hatch with a sigh of relief. We are nearly weightless and pleasantly cool, although aware that the chill will soon pervade our bones and turn the final minutes of our dive into agony. We try not to think of this as we go through the meticulous routine of checking each other's gear for flaws, leaks, or possible malfunction. Check that our weight belts are locked against accidental release; then ensure the smooth function of exhaust pop-off; normal bypass operations; open control block (remember Tiger Manning's near death); and, finally, proper buoyancy, with all safety gear checked in place. Now off to the psychologist's workbench, alongside the shark cage. First, the strength test; next, the two-handed coordinator. But watch! There are two big scorpion fish on top of the gadget; no stings today, please. And so through the triangle assembly—to what purpose, we wonder?

After 15 minutes of impatient fiddling with these devices, we are free to start our journey. Down to 1,800 pounds of gas now, we must save enough to repeat the tests, plus 600 pounds reserve. We start off, hooked together by a buddy line, following the polypropylene strand that shines in the lead diver's light. Visibility is reasonably good—about 6 feet. The mud bottom is studded with hundreds, even thousands, of scorpion fish, their poised hackles erect and waiting a careless touch; elsewhere, schools of anchovies, croakers, and occasional cabezon swim with us, further obscuring our meager light.

With every kick of our flippers, a dense cloud is raised; if we stopped, our visibility would be reduced immediately to

6 inches or less. But we are not stopping today; approval has been granted for a sightseeing tour to the canyon. We take a bearing on an electric ray lying beside our last way station, veer to the right, and glide down an irregular 40-degree slope. Two hundred forty feet down now, and nothing but gray mud and drab bottom life. Then, abruptly, something happens to the water ahead. Now the dark layer is overhead, and for about 20 feet above the bottom the water is almost crystal clear. It appears that all the light is coming from the sea floor, while the overhead layer is black and intimidating. It is a strange reversal of the usual scene, but comforting, as if we are now in the strange zone of life, light, and safe water, while the environment above is our enemy. We swim ahead to meet this friendly stratum of existence.

Abruptly we are at the brink of the canyon, probing with our lights over the precipice and startled at the sudden increase in visibility, up to 40 feet. Strange to describe a clear blackness, but that's the only word for it. Now, we see immense boulders, densely covered with plant life of all kinds and of infinite beauty. Gorgonian fan corals wave in the slight current, and a world of color comes alive under our lights. This is the first time we have seen living plant colors since leaving shore. Around us, the variety of fish is amazing, and as far as our vision can probe along the canyon rim and wall, new and beautiful forms of life and rock appear. Far below us, the canyon floor is narrow and unseen, perhaps to be probed by men in our lifetime, but not on this trip, nor by this pair.

Our depth gauges, forgotten for a moment, read 268 feet. Watches show 20 minutes of elapsed time since leaving the psychologist's gadgets. Our time is up long before our visual search is half begun. We drive marker stakes, look over our shoulders, and start back along the trail of nothing, leading nowhere. We swim quickly back to the work platform, to grind through the tests, teeth chattering on our mouthpieces, and shivering uncontrollably with the bitter cold we did not feel at the canyon's edge. Just finish the damned tests and head for those hot showers.

Later, as we slowly warm, we think again of the empty black vistas that separate Sealab II from a beautiful spot on the ocean floor. This has to be learning undersea geography the hard way. Now two other aquanauts depart to check the weather station. We instantly feel sorry for them; their entire day will not equal the few minutes we had on the verge of Scripps Canyon.

The preceding narrative is a big part of the story of Sealab II, and I hate to have to write it as seen through eyes other than my own. The day is now ended for me, and the tour ended for ten aquanauts, with only a passing glance at those things on the ocean bottom that make being there worthwhile.

26 September 1965

This has been another day of truth for aquanauts and top-siders alike; a good day, a happy one, even hilarious at times. Early today the press and TV flowed over the gunwales of the *AVR-10* in a seemingly endless wave of faces and forms, to take up their perches in dangerous spots on the *Berkone*, daring the sweep of a headache ball or pennant and shackle as they observed and recorded the transfer of aquanaut teams 2 and 3. From the start, the transfer moved smoothly, with no hitches discernible to this practiced eye. Over the television monitor, we watched the ten aquanauts of team 2, one at a time, duck through the exit hatch of Sealab II and, holding their breath, swim over to the personnel transfer capsule, reporting promptly on arrival. The capsule was buttoned up, and the rise to the surface got underway.

Overhead, a helicopter hovered, blasting the visitors with salt spray and drowning out all commands issued by the boss rigger. The PTC finally emerged from the water, was deposited in its cradle, dropped ballast, and mated smoothly with the pressurized deck decompression chamber, all in an amazing 26 minutes. Minor difficulties developed at the point of disconnecting the two, but these were resolved by inadvertently blowing the O ring far out into the Pacific. No casualties were noted among the spectators, who thought the episode to be a scheduled part of the show.

Soon the PTC had again descended safely to a location outside the shark cage, and three aquanauts of team 3, who had stayed overnight in the habitat, requested a 4-hour respite from visitors, to permit further housecleaning and badly needed replenishment of expendables, victuals, and life support hardware. During the wait, the remaining team members stretched out in the sunlight, from which they would soon be hidden for more than a fortnight.

My private unlisted telephone rang about noon, with a call from Mr. Jack Valenti of the White House staff to inform me that President Lyndon B. Johnson would like to talk with Scott Carpenter by phone this afternoon to express his interest in the Sealab program and Scott's role therein. After a period of understandable confusion, we worked out a telephone patch with Scott, who was confined with his team in the deck decompression chamber. As I awaited the important call, I savored to the fullest my part of interlocutor in the exchange.

When my telephone rang, I sprang into action. Like the conductor of a great orchestra, I bade the tumult about the deck to cease, signaled to start a tape recorder, and brought Scott Carpenter into play, all with a single gesture and some button-pushing. Having introduced myself to the two telephone operators involved, I suggested that one of them speak directly with Commander Carpenter while I modestly lingered in the wings to cue either party, should lines be forgot.

The local operator opened with the innocent question,

"Comdr. Gordon Cooper, astronaut, would you speak a few test phrases to check our circuit?" The reply, typical of Scott's intellectually fastidious approach, went about as follows.

"Negative! This is Scott Carpenter, aquanaut. Astronaut Cooper was last heard of in Tanganyika; would a test phrase in Swahili do as well?" The balance of his opening remarks, conveyed in pure helium speech, seemed to harbor some painfully explicit directives delivered over an extremely garbled circuit. Then came the warm, friendly voice of Mr. President, and the sun shone clear on blue waters. The conversation went superbly, although the non sequiturs bore testimony to my suspicions that neither party was able to grasp the trend of thought at the circuit's other end. However, since I am perhaps the only man who could interpret what was said, considering my southern background and philological agility in helium speech, I intend to withold further comment until the heat of the next campaign.

28 September 1965

In the early hours today, I prepared to cope with the onslaught of press and dignitaries who would soon arrive to witness the exit of Scott Carpenter and his team from the DDC.

The early arrivals appeared and problems of camera locations and priorities began to arise. Word got around that *Life* magazine had clandestinely installed cameras in the DDC, and I was soon harassed by rival photographers, both amateur and professional, for equal treatment. This involved repetitive operation of the medical lock for equipment transfer, which, in turn, meant interruption of the linear decompression. By noon, our exit time had slipped an hour and I was getting irritable.

About this time the chamber corpsman, Ray Lavoie, cornered me near the trailer to convey the disquieting news that

Scott was complaining of pain in both knees, at a chamber depth of 44 feet. Walt and I decided to go in for a look-see. Within the pressure lock, Scott lay on a bunk, somewhat discomforted but not in severe pain. His complaints were of cramps in both quadriceps muscle groups, with diffuse, unchanging pain in both knees, not localized, and worsened by activity. Physical examination revealed nothing of note, but on questioning him I learned that he had spent more than an hour in an acute yoga sitting position on a bunk, and that the pain had commenced thereafter, some hours ago. I was certain that the cause of the pain was not the classical bends, basing my diagnosis largely on the fact that both knee joints had been involved simultaneously, plus the fact that Scott had come nearly 45 feet nearer the surface with little if any change in the intensity of his symptoms. Under the circumstances, such a rise would ordinarily have compeltely incapacitated a bends victim, and would have resulted in additional "hits" in other joints, or in the spinal cord. Content with my own reasoning, I ordered a mild painkiller, some activity, and a resumption of decompression. Walt and I locked out of the chamber, to resume our separate watches. Within minutes, I was queried by newsmen anxious about the bends in Scott Carpenter's knees.

When at last the chamber hatch opened, the aquanauts of team 2 emerged to cluster about the entrance for group photos. Two wives were present to kiss their spouses, a few words were spoken for the record, and the men were off to the showers. Some minutes later, we were medicating both Scott Carpenter and Howie Buckner, the former for general leg cramps and pains and the latter for muscle strains resulting from chamber acrobatics. Although I was unable to convince Scott that his pains were not a case of neglected bends, Howie accepted his lot without complaint and trudged to bed with his pills. Scott required further medication during the night and early morning, but seemed improved as he walked the length of the pier and up the inclined road to the auditorium at Scripps Institution for the scheduled press conference.

I had been required to write out my opening remarks so that they could be printed as handouts for the press conference, where I would read them. In something over thirty years at lectern and pulpit, I have never written a line in advance, and all printed reproductions of my narrations have been after the fact. Nevertheless, this day was to be different and not at all to my liking. Late the night before, I had reluctantly dictated my piece over the phone, with dire admonitions that spelling must be accurate. Considering that I have never seen a naval document of more than three paragraphs without a misspelling, I quailed at the thought of orthographic errors going out under my name.

The conference went along nicely, with all the aquanauts giving good accounts of themselves, ending with Scott Carpenter, who spoke articulately and to the point. Scott briefly voiced his opinion on the need for more autonomy for the sea-bottom dwellers. Recalling some of the events of the past thirty days, I could not help but remark to myself that more autonomy would have been fatal to the lot of them. The degree of autonomy in Sealab II, a new and unsophisticated program, was at least an order of magnitude greater than that available to the space twins of the Gemini Program, years older and infinitely richer than our own.

An hour later, Walt and I fled the scene to return to the quiet of the *Berkone*. Even here it has not been so quiet. Dr. Bob Sonnenburg called up from Sealab II to say that he had a reading of 200 parts per million (ppm) of carbon monoxide on his test kit—a shocking though unlikely bit of news. Unlikely, because the men were still well; shocking because of the possibility that we have a carbon monoxide (CO) producer, aside from man himself, inside the habitat. I ran a hasty CO test on our topside kit, and got less than 100 ppm, still far too high but, considering the crudeness of my test procedure, probably far higher than the true value. A repeated, more meticulous test yielded a level of only 20–30 ppm. That value is compatible, even at 7 atmospheres of pressure, with good health. Nevertheless, we immediately took the following steps: a gas sample was dispatched to Linde Laboratories for analytical study; aquanaut blood samples were dispatched to a special technician at USNHS for carboxyhemoglobin studies; and Hopcalite was procured for incorporation in the Sealab II scrubber system.

I am sure that the high CO reading, when reported, will cause a flurry in Washington, largely because of the conflicting opinions on the subject. Much of the disagreement was threshed out in our committee on hyperbaric oxygen months before the advent of Sealab II. In fact, carbon monoxide effect on mammals under pressure is not proportional to the pressure, yet it is somewhat greater than for similar exposure at sea level. And that, by God, is all that is definitely known at present!

Thirty days has September, and this is one month I'll remember.

1 October 1965

Apparently within minutes of receipt of my SitRep, persons were seeking out the experts, and I was shortly receiving all manner of gratuitous advice regarding safe or permissible levels of CO for prolonged human exposure at 7 atmospheres of pressure. As the problem of CO under high pressure has been carefully considered by our own experts on hyperbaric oxygen, and as no one has yet done and reported the laboratory work required, I was not wholly appreciative of this advice. I believe in experts, and use them to the fullest extent practicable, but in the practice of medicine I choose my own.

Down below, the aquanauts began their work in a series

of salvage-related tasks assigned by Capt. Willard F. Searle, Jr., supervisor of salvage. Of high priority in this portion of the program was the use of a foam-in-place technique to raise a sunken object, in this case a jet aircraft hulk, from the ocean floor. The aquanauts were to insert the nozzle of a squirt gun into a compartment of the aircraft, then squirt a premixed solution into the compartment space. Ideally, the solution immediately forms a buoyant foam that adheres to the inner walls of the object to be salvaged, then solidifies into a rigid sponge, giving sufficient buoyancy to raise the sunken object. The concept is new, expensive, apparently quite safe, and ideally suited for an operation such as this one.

In actual practice, as is so often the case, things did not go so smoothly on the ocean floor. Early in the foaming phase, the beautiful, expensive material began to flow out of hitherto unseen cracks in the fuselage, limiting and then completely obscuring the divers' vision. The adhesive characteristics of the foam particles then started to live up to full, unforeseen potential. Gradually the divers' wet suits built up a thickening layer of the white stuff until they were almost totally covered. The purpose of the foam is to increase the buoyancy of the object to which it adheres, and aquanauts were no exception. Since buoyancy is synonymous with certain death for the saturated aquanaut, divers Bob Sheats and Bill Meeks beat a hasty retreat to the habitat, struggling to stay near the bottom. About half-way back came the final blow. Like glue, the foam fastened itself onto Meeks's exhaust valve and sealed it firmly shut. The pair limped home one flipper's length ahead of fatal disaster.

After falling back and regrouping, Sheats's indomitable workers attacked the salvage assignment with new respect. The foaming was successfully completed, using double 90s instead of MK VI breathing gear, and the initial trials of explosive-driven stud guns and cable cutters were completed.

The foam is efficient and will undoubtedly prove to be valuable as an alternate salvage technique, but we must learn how to foam-proof the men who use it. Back in the habitat, the aquanauts rested and prepared for a late-night sortie to Scripps Canyon while I dreamed of going to another canyon—the Linville Gorge, in North Carolina. It was a nice trip and a sound sleep.

2 October 1965

Early this morning, Walt and I watched and photographed a beautiful sunrise. This was to be the day of communication between our aquanauts and the oceanauts of Jacques Cousteau, separated by some 7,000 miles of land and water. Cousteau's Conshelf III operation has a six-man crew in a habitat at 99 meters off Cap Ferrat. Scheduled for a fourteen-day bottom stay, the oceanauts were in their tenth day of the project. After considerable difficulty, a telephone hookup between the two habitats was established. The voice exchange would not only be bilingual—in French and English—but would also originate in environments of ordinary air, helium-oxygen, and Conshelf's neon-oxygen. Given this mix of atmospheres, the potential for successful communication was staggering.

About an hour before the hookup, I made my second brief trip to the beach in fifty days. On touching ground, I experienced the disagreeable rolling motion of the pier and terra firma around the Scripps campus that I had noticed in the early minutes of my first visit ashore. Not wishing to embarrass my hosts, who seemed to find nothing amiss, I fought the urge to call attention to this unsettled condition of the land mass despite my personal conviction that a major earthquake was occurring directly under our feet. The unpleasant state lasted until our arrival at the Sealab II Command Information Bureau provided a welcome distraction. We assembled briefly, then departed for the bomb shelter that harbors the nerve center of the Benthic Lab.

There, bearded and sandaled graduate students sat silently before the TV monitors, studiously watching our aquanauts and logging butt scratches and other motor activity. In a dark corner, a professor with a preoccupied air was pacing in circles as he listened to tapes of the previous night's conversation inside Sealab II. On occasion, he would pause and scribble furiously on a nearby bulkhead as a fanciful anecdote of bedroom lore issued from the speaker. I ventured to suggest that more dramatic stories might be found on the walls of Grand Central Station men's room, but he silenced me with a stare.

Promptly at 0930 came the call from Conshelf III, with André Laban on the line. Speaking a few words to Al O'Neal, André was off to a fresh and original start, but the conversation fell into the doldrums of banality very soon. I seized the instrument from Al and greeted André in ringing tones of Swiss French. André delivered himself of some choice phrases that took me back to my days in the gutters of Paris. At this point, I made an important discovery. In French, as in our tongue, the classic but socially unacceptable words will break the helium barrier with ease. Gradually the conversation sorted itself out and I relinquished the phone to Scott Carpenter, who thought he was talking to Cousteau's son, Philippe. André, still on the line, thought he was again addressing O'Neal, since the language had reverted to English. Then Philippe came on the line, speaking excellent English heavily flavored with neon. At length it was established that Scott was happy to talk with Philippe and, by remarkable coincidence, Philippe was gratified and happy to speak with Scott.

Later, comparing notes for a press release, we summarized the following informational exchange: The oceanauts and aquanauts were well; the oceanauts congratulated the aquanauts; the aquanauts congratulated the oceanauts; the water in the Pacific was dark and cold; the water in the Mediterranean was dark and cold; we did not understand each other very well. What wasn't expressed in either language, but was clearly evident to those of us online, is that a strong bond of fellowship exists amongst all who share the experience of living underwater.

3 October 1965

It appears that team 3 has generally been the happiest of all aquanaut groups. A sense of good humor prevails, which may be a function of team composition, but more likely is related to accomplishment of the daily schedule. Certainly it is the easiest team to work with from topside. A little conversation several times daily and all jobs done with no hassle and a minimum of detail-threshing. The upcoming work of the day having been dispensed with, we generally lapse into an unhurried discussion of life below the surface. Sometimes we are led on a fantastic journey across the early morning seascape seen through the eyes of television and narrated by an observant aquanaut.

This morning, as we talked with Bob Sheats, he casually asked if we would care to see the aurora borealis of the ocean atmosphere. Unbelieving, we scoffed an assent as he darkened the lights in Sealab II and directed our attention to the large porthole. In a few moments, at the upper corner appeared a swirling mass of light that resembled the spiral nebulae as illustrated by astronomers. As the light grew in size and intensity to fill more than half of the port's vista, Bob described its luminous qualities and peculiar patterns. Pointing out other features of the undersea environment, he spoke of eating a handful of euphausids the day before and finding their flavor excellent. Plans were made for a plankton soup dinner later in the day. Being a forager by disposition and training, Bob could live off the resources found on the ocean floor as readily as he survived years ago in a Japanese prison camp. It strikes me that undersea living in the total sense calls for a great deal of woodsmanship, which is not easily imparted to a person who does not truly love natural phenomena.

Soon came time for our weekly Church of the Sea services, which I truly enjoy and which have become a part of our way of life in the program. After an abbreviated service of morning prayer, I talked about the neglected value of emotions in making useful judgments and also in arriving at points of personal philosophy. Feeling rather strongly on this score, I am sure that I rambled a little. Perhaps it is unfair to use these minutes as a sounding board for my personal convictions and philosophy, but if it gives the aquanauts better insight into the motivation of Papa Topside, then the time is well served.

The balance of this pleasant day went smoothly, with more excursion dives and marine biological studies. Despite the cold water, the efforts of the past workweek, and the difficulties of dressing, water entrance, and egress, our aquanauts piled up nearly 500 minutes of bottom work time. Later in the day, Bob reported success in making pancakes, a feat that has eluded all previous underwater chefs. The secret lies in making an ultrathin batter.

Topside, we drew a long breath for the stretch run.

4 October 1965

This day came dangerously close to being routine, if such can ever be said of sea-floor living. After checking the weather report and having a leisurely chat with Bob Sheats, Walt and I took time to reexamine the haunting question of atmospheric contaminants under high pressure.

Maintenance of a safe and breathable gas mixture inside any completely closed human habitat will always be the most critical aspect of a venture, whether it relates to spacecraft or to undersea houses. In space travel, however, the permissible margin of error is reasonably broad since, in this case, a swing of 5 percent in oxygen content is quite tolerable and CO_2 levels of even 6 percent will not be hazardous to the astronaut. In high-pressure living below the waves, however, this luxurious latitude is not permissible. At a depth of 600 feet, oxygen levels must be maintained at 1 percent, with virtually no margin for error, and carbon dioxide must be scrubbed down to about 0.06 percent. At depths greater than this, oxygen, the breath of life, becomes no more than a trace contaminant, and tolerable levels of CO_2 are lower than those found in pure country air!

Of added concern in any pressure experiment is the matter of trace toxic contaminants in the atmosphere. The toxic substances, largely derived from the family of volatile hydrocarbons, are a source of deep concern to any investigator. Their presence in the captive atmosphere is inevitable, and the variety and numbers of these invaders are incredible. A classic example of this may be found in a recent Air Force report of a fifty-six–day chamber exposure in which heroic measures were taken to exclude atmospheric contaminants. Near the conclusion of this run, gas sampling revealed the presence of about 170 contaminant gases, of which more than half were toxic to human beings, and 10 percent of the remainder were of suspected but undetermined toxicity.

The sources of these dangerous contaminants are varied and, in some cases, completely mystifying. The occupants of the habitat contribute a fairly large share of the metabolic castoffs; nearly every fabricated item in the habitat can be counted on to yield a generous share; and many of the seemingly inoffensive pieces of equipment, such as electrical motors and relays, will add to the toxic burden.

To the Man-in-the-Sea investigator, this poses a horrendous problem, since identification of these toxic contaminants is a tedious and imperfect art, and elimination is only partly possible. The final and most severe problem is that literally nothing is known about toxicity increase at elevated pressure. It is possible, even probable, that the toxicity of many of these gaseous compounds is a function of their partial pressure, rather than their relative concentration. We have no current answers to the problem, and it is distressing to contemplate that a single pressure-toxicology complex, costing more than $1 million, could not give all of the answers within our lifetime or that of the next generation. Obviously, we approach this problem with a philosophy of calculated risk.

Such was the tenor of our talk as the sun burned through the fog, the *Berkone* came to life, and our aquanauts below prepared themselves for another assault on our friendly adversary, the sea. As the day wore on, team 3 continued to perform a variety of salvage tasks with skill and accuracy, unshielded divers successfully used explosive penetration devices, and exploratory canyon dives were made to 305 feet. Three separate scorpion fish stings were inflicted on Bill Meeks, who responded rapidly to therapy, and I lost a front pivot tooth. As dental care is not provided at sea, I will unsmilingly endure this tragicomedy until my next excursion ashore.

6–11 October 1965

As we come to the end of this 45-day experiment, I am impressed with the efficient routine of each 24-hour span. By now we know almost precisely how long it takes to do a given task and how much work will be accomplished on a given day. Although it is gratifying to note this functional teamwork, it is disturbing to reflect on the time required to effect the transition from the performance of the first team. This is understandable when we consider that the aquanauts of team 1 were physically and emotionally fatigued from the strain of the last uncertain days of outfitting and placing the habitat on the sea floor. Once actually inside Sealab, they spent the first week in continuous, frantic activity aimed at setting up all the paraphernalia necessary to safe and productive life below. Psychologically and physically for all the aquanauts, their time spent below was an abrupt exposure to an environment more hostile and demanding than ever before experienced by divers, and this was the case every day, seven days a week.

Furthermore, the topside control team was equally beset with fatigue, fear, and equipment failures at the beginning. Anxiety for the safety and well-being of the team 1 aquanauts resulted in far too much chatter on the overburdened communications circuits, and many of the orders from topside must have seemed whimsical and arbitrary, if not downright sadistic. During this shakedown period, there were inevitably small equipment failures, both topside and in the habitat. As breakdowns had been anticipated, redundancy of systems and elements was liberally provided by the fabricators so that repairs or replacements could be done on company time. But topside, we felt that any hardware failure, however minor, was a threat to safety. Immediate correction of any deficiency was therefore the order of the day—and night as well. The personal discomfort of the aquanauts and the disruption of their routine received little consideration. The repairs or readjustments must always be done at once.

A final disturbing factor of some importance was the difficulty in integrating the many projects planned for simultaneous operation, each project being accompanied by its own cadre of undersea and topside personnel. Although a rank order of priority had been assigned beforehand, plans tended to fall through the cracks for many of the projects, resulting in partial or total failure.

The root of this problem was the lack of time for a thorough checkout of the entire sequence of operation in shallow water prior to the deep run. If time had permitted, we could have debugged the entire complex and identified and

A night view out the Sealab port. The array of devices includes atmosphere monitoring equipment and a centrifuge for blood studies. (U.S. Navy photo)

discarded those projects that yielded spurious data or none at all. So much for hindsight; we have learned that an experiment that requires more than a year of preparation and costs nearly $2 million can afford a few weeks of proper shakedown, under realistic environmental conditions.

12 October 1965

The job is all but done. Four hundred fifty man-days of life and work on the ocean bottom have been accomplished without serious accident. More than a half million items of specific information have been punched onto cards or processed through the undersea multiplexer for direct computer analysis. Correlation and cross correlation of these data will be in process for months to come. From these data will come our guidelines for extension of human exploitation of the continental shelves or beyond, perhaps even to the abyssal plains.

To me, this is a story of brave and adventurous divers—a new breed we call aquanauts. In the case of our military personnel, I have dived and worked with them for almost seven years. Our civilian volunteers, scientists and engineers, were aquanauts from inception and team members from the start. My theory is that only the best will volunteer for the program in this early and experimental phase. Well, the days have passed gainfully, and I am deeply satisfied, as are all others who stood the topside vigil. Now comes our last transfer and decompression of saturated divers; as always, I fear this phase of our operation.

Transfer of personnel from the habitat to the decompression chamber is always a dangerous procedure. The routine sounds simple enough. The aquanauts swim freely from Sealab to the nearby personnel transfer capsule, enter, and after a careful systems checkout, close the lower hatch and prepare for liftoff. From this time forward, any leakage of valves or exposed gas lines in the PTC will result in almost immediate death of all ten occupants. Considering the inevitable function of Murphy's Law of the Sea, I view these transfers with horror. Yet they must be done, albeit with antiquated handling methods.

This morning, as we began our final PTC liftoff, a journalist turned to me and said, "Well, Captain, I guess you've got yourself a no-hitter!" In our parlance, a "hit" is a diving accident, be it embolism or bends. I shuddered and replied, "Don't say that until the last out!"

The PTC came clear of the water and was swung across to the area where the ballast would be dropped. The sea state was nearly zero; nevertheless, the pendulum action of the 13-ton capsule was frightening. After ballast drop, we swept the capsule and its inhabitants over to the mating area of the deck decompression chamber. With some difficulty, the mating process was completed and our aquanauts transferred to the relative safety of the large chamber, still under high pressure. The tedious process of decompression began aboard the support ship, *Berkone*.

In this last decompression, Walt and I had agreed on a total time of 33 hours, since we had anticipated several stops that would extend our schedule. By and large, all went well for the first 24 hours, but while approaching a depth of about 60 feet, trouble struck. Team leader Bob Sheats, age fifty, called me on the intercom sometime after midnight with the unhappy information that he had developed bends. I locked into the chamber to verify the diagnosis, although I knew that Bob Sheats would not be wrong. He was indeed bent; he had a severe central nervous system bend that affected his right leg.

A decision was in order. Nine of the ten men in the chamber were within a few hours of release. It might be necessary, however, to recompress Bob Sheats to 500 feet or more to treat his bends. If I elected to place the other aquanauts in the outer lock of the chamber for their final hours of decompression, Chief Sheats would be isolated in the inner lock, and could not be reached for medical assistance. Otherwise, the entire team would have to recompress along with Sheats, then undergo nearly another week of decompression.

I made the decision. Tiger Manning, our highly competent medical diving technician, locked in with Bob Sheats, and the remaining nine aquanauts were transferred to the outer lock, their decompression to be continued by Walt. Under Walt's meticulous care, the nine were safely decompressed in less than 3 hours' time. I undertook treatment of Bob Sheats. By the grace of God, and tables prepared from the wisdom of Capt. Robert Workman, Bob Sheats was cured of the bends in 6 hours. At long last, the experiment ended safely for all.

There remained only a press session at Scripps Institution, which lasted for more than 2 hours, but we lived through it. As I looked at the aquanauts, I felt overwhelmingly proud of them. It is very possible that we will never see their like again in our time. I have been deeply honored to have such a team, and I love them, one and all.

18 October 1965

Now the trumpets of the heralds are muted, and the work of packing up for the next stop is in the hands of the aquanauts. For a brief five days after the final press conference and ceremony, I made a point of traveling fast and incognito, and so regained a semblance of sanity. I was a nomad, stopping at unannounced spots and sometimes staying the night at a hostelry whose name I never bothered to learn. It was my personal approach to unraveling. Once every day, from a pay telephone or motel room, I called in to verify my existence and to assure myself that the world turned without my guidance. Almost every day, I touched base with some activity within the scope of the Man-in-the-Sea program, but always as a bird flying, with no point of departure and no forwarding address. For five great days the world was kind, allowing nothing serious to happen.

Although alcohol was always at hand, I drank very little during those days of renewal. I spent the time scanning new vistas, sleeping prodigiously, looking at the female face and form, and visiting several Navy laboratories and activities. For 2 hours in Golden Gate Park in San Francisco, I watched the animals and savored the green grass and trees.

Then I drove back and forth along the Pacific beaches and explored the lesser ranges of hills to the east. The clock unwound smoothly, and I returned to the starting point of this Chronicle—the Outrigger Inn, Long Beach, California. My wake-up call was for first light, but Washington, D.C., marches to a different time. Word was out that Sealab II had been successfully completed, and it was assumed that I was available. First came the expected requests from organizations to which I owed talks long since postponed. Of greater consequence was a call from my home office in Washington with an invitation for a command performance at the Pentagon, this of all weeks! I tried desperately to forestall the inevitable, but to no avail. Bob Sheats, Scott Carpenter, Walt Mazzone, and I were required to appear at a Pentagon "nooner," Thursday, 21 October, under an official invitation from the Secretary of the Navy himself. We would, of course, comply, although the invitation created problems. At my home in Washington, I had the required blue uniform, and Walt was suitably equipped in California, but Bob Sheats had to fly to Seattle for his, and somehow I had to locate Scott and divert him to the new appointment. Since both Bob and Scott were to receive the Legion of Merit from the hand of the Secretary of the Navy himself, the occasion was of some importance.

By virtue of countless long-distance phone calls and several trips to the shipyard, all parties were informed at last and ready for assembly. At the Pentagon, the ceremony and news conference were dignified and well managed. Scott, unfortunately, could not appear, being grounded by fog in Hartford, Connecticut.

At long last, we were free to board a return flight to Los Angeles to finish wrapping up Sealab II. Once we were airborne and had settled into the monotony of the flight, a lulling sense of tranquillity came over me. Somewhere over Colorado, the opening lines of a poem came to me: "I shall not pass this way again/ This way I loved so well!"

8

Sealab III
(1967–1968)

It was a sad day on 27 January 1967, when three pioneers lost their lives in a fiery holocaust of mercifully brief duration, in a spacecraft that represented the state of the art of this nation. Three astronauts were killed, not in the turntable of weightless spin and speed in outer space, but in a single moment of incineration atop a tremendously complex launch pad. Our nation is shocked, grieving, and stunned.

Of the three victims, I knew one well, if that is an adequate description of friendship between people who meet in harm's way to prove a point of science. I knew Ed White, age thirty-six, as well as two individual adventurers can know one another, however far apart their fields of endeavor might be. We first met at Quonset, Rhode Island, where he was pilot of the aircraft that took my diving tank team on a series of Keplerian trajectory flights identical to those completed by the Mercury astronauts a few weeks before. Ed White returned faithfully to pick up new crews until all of us had been exposed to nine periods of quarter-minute weightlessness, both inside the "black box" and outside as cabin passengers. We wrapped up the weightless flights about 2100 hours in the evening and got together for an hour of debriefing. Ed White was particularly curious about our comparison of underwater weightlessness with the gravity-free state that we had all briefly experienced that day.

As a result, aircraft crew and aquanauts alike boarded a bus to the 118-foot escape training tank, 60 miles to the west, in New London, Connecticut. At midnight, we opened the tank, lighted the clear water, and began to train our friends in the problems of weightless operation, this time in a medium eight hundred times denser than air. After 4 hours, a great deal of knowledge had been exchanged, mutual respect had been established, and a lasting friendship had been struck among pioneers of the water and those of the air. As we parted company at dawn, we promised to meet again.

Some years later, after the Mercury program was history, Scott Carpenter invited me to an astronauts' Christmas party in Houston. John Glenn, recently recovered from his back injury, drove the group of us to a hotel where we met the newly appointed Apollo I astronauts. One of them was Ed White, and the two of us exchanged shared memories.

In a temporal sense, we lost three dedicated men on that day in January 1967. It was no comfort to three widows and their families that a lesson had been learned. Nor was this the last time such a tragedy is likely to occur, in either outer or inner space. But the quest, and the progress, will continue because of people like Ed White, pilot of Gemini 4 and the first American to walk in space; Gus Grissom, one of the seven Mercury astronauts; and Roger Chaffee, who died before he could fly in space.

By the time Sealab III was conceived and tentatively proposed for the summer of 1967, I had become an observer, chronicler, and occasional adviser to the project. I had suffered through a series of operations during the 1967 winter months for a condition that started with a crick in the back involving my right leg and escalated to partial paralysis of my hindquarters, accompanied by unremitting pain. I first submitted to a myelogram, an operation in which a needle is inserted within the wrapping of the spinal cord and the patient is tilted, twisted, and probed under the fluoroscope and X-ray until the last mysteries of the cord derangement are clarified. I then readily consented to examination in the amphitheater of the National Naval Medical Center by a visiting civilian orthopedic surgeon of Scottish heritage and considerable reputation. On a gurney stretcher I was wheeled down to the lower areas of the hospital, where I was told in a Lowland Scots brogue to crawl off the stretcher, stand up straight, and drop my pajama bottoms. I did the best I could but required assistance, and stood for an agonizing 2 minutes with my face to the wall, listening to the description of my crooked back, uneven buttocks, and complicated medical history being presented to students, interns, residents, and especially to the talented specialists who had assembled to hear the final diagnosis. At long last, the consultant and others concluded that I should have my backbone welded in a surgical procedure designed to relieve my ailments. When I asked the obvious question, I was firmly advised that I would never dive again.

During the long 1967 spring of convalescence and therapy that followed, I managed at least to lower myself into a pool and work up to a 700-yard daily swim. I was greatly encouraged by the example of the paraplegics and Vietnam casualties, in worse shape than I, who worked out in the pool with me. Despite the almost complete success of the rehabilitation program, I vowed I would never place myself in the hands of my colleagues in the medical profession again.

When I returned to the Sealab project in July, I found that the characters and staging of the drama had shifted, but the basic plot was remarkably unchanged. Five teams, each made up of nine Navy and civilian aquanauts, were to live in the habitat at a depth of about 430 feet for twelve-day periods. The habitat, a reconfiguration of Sealab II, would be anchored off San Clemente Island, California. The Man-in-the-Sea program had finally caught the attention of many supportive Americans.

The Sealab projects were a popular element of the renamed Deep Submergence Systems Program, but of the thousand-odd people involved in the new experiment, only two naval officers had previous experience with the habitat and with saturation diving procedures—Walt Mazzone and I. We were vastly outnumbered by well-meaning persons who had acquired sudden intelligence of matters that had taken Walt and me our entire naval careers to develop. I was startled to find that the records of previous Sealabs, meticulously maintained, were generally ignored. It is difficult to pinpoint a given source of incompetence, but Sealab III, scheduled for completion in 1966, was stalled in its tracks. The target date became a distant, rather than an immediate, goal.

While the personnel problem was difficult, it was not without a solution. Clearly I could not singlehandedly run the show; someone must be brought aboard to operate our remote aquanaut facility, and a line officer was needed as the central project manager. The choice of the operator was obvious: Walt Mazzone was the only one with the experience and zeal for the job. For the Washington-based post of project manager, I selected Comdr. Jack Tomsky, a former submarine rescue skipper with a reputation for being hard-nosed and for accomplishing his missions. Comdr. Russ Drake, my administrative officer, was already aboard. There were billets for Jack Tomsky and Walt but not for the other aquanauts— enlisted men, chief petty officers, and officers—necessary for the program. Jack Tomsky took this problem in hand and was finally successful in acquiring some fifty-odd billets by application of refined techniques of extortion not recognized in naval regulations.

While these intrigues were afoot, it was necessary to find a home port for our technical office. After a six-week search for a suitable geographic location, the site of unanimous choice was Ballast Point, San Diego, California.

In September 1968 I finally bade farewell to my family for

a separation that would last until the third adventure in undersea living was completed. For better or worse, and I hoped for the last time, our family umbilical strands were again broken. I reminded myself that most of our aquanauts had been away from their families for a year by now, while I would be absent for only a few months.

A week after arriving at San Francisco, my first stop en route to Long Beach, I was laid low by a vicious sore throat. I resigned myself to bed rest in my motel, a diet of wonder drugs, and paperwork. That was where Paul Linaweaver found me to show me the X-ray evidence of bone trouble in the cases of Scott Carpenter and Joe MacInnis. In Scott, the films showed clear-cut lesions a few inches above the knees, both port and starboard, corresponding all too well to the areas about which he had complained during decompression almost three years ago—and which I had dismissed as inconsequential muscle disorders. The X-rays for Joe were less distressing, showing only a single discrepancy in one leg, which did not resemble any case of bone disease.

I had planned to put Scott on the bottom with one team, with the same possibility for Joe. Given the X-rays, however, perhaps neither aquanaut should live in the habitat during this exercise. After long thought, I decided that Scott should not again be exposed to a saturation dive and subsequent decompression, while Joe could remain on the list of candidates. That was my own judgment, subject to reversal after review by a select group of radiologists. Regardless, it was quite possible that I had misdiagnosed Scott's case some years before. Prior to our program of saturation diving, we were instructed that bends were never bilateral and never related to muscle pain such as Scott experienced. But we had since seen more occurrences of such muscle pain, which might have reflected a bubble block of an end artery within the bone. The question of end arteries looked like a critical area of investigation for the future, considering the similar effects seen in the structure of the inner ear, leading to the Organ of Corti. We had seen no fewer than eleven cases of impaired hearing during decompression of saturated divers in the previous twenty-two months. The problem needed to be investigated.

By the end of September, I began to sense that most of our team were tired beyond belief. All, I think, were ready to move on from San Francisco to the operational areas of Long Beach and San Clemente. We had been too long delayed by unrealistic planning, constantly revised documentation, and arbitrary changes that were not always passed on to the three hundred–odd personnel. An unfortunate example of this problem was the case of the upside-down underwater davit as a means of handling the pressurized pot that served as dumbwaiter for the transfer of materials to and from the habitat. The basic design was nearly perfect: it did not require a diver in the water, and no mule-hauling was needed. The pot was delivered hands-off to the habitat and returned to the surface in like manner. However satisfactory it may have been, someone attempted to improve the engineering by designing a hoist to bring the pot to the surface at a fantastic rate, venting to equal ambient pressure all the way up, with almost no manual control.

For bringing up a handful of notebooks or other nonviable objects, the revamped equipment might have done nicely. But a critical portion of the program was transport of biological specimens, which, unlike cardboard objects, had to be adequately decompressed. Decompression schedules for biological samples had been established for Sealabs I and II during test runs at the Experimental Diving Unit, but this information had not been heeded, evidently because *no one had read the record.* As a result of this misconnection between physiological investigators and design engineers, the delivery system did not meet the minimal needs of the biomedical program. Had we been consulted on design, a change in the ascent rate of the transfer pot or the addition of an external gauge with a pressure release valve would have solved the problem. Furthermore, as the pot would not fit into the medical lock of the deck decompression chamber, we had to rethink the transfer methods planned for the many blood samples from the aquanaut occupants of the DDC.

By early October, as I continued to check on test procedures at the shipyard, I found that an old truth had been rediscovered: Helium is an elusive and pervasive gas that will leak into and out of any enclosed cavity. Currently, the best containment theory calls for metal-to-metal seals with a minimum of gaskets. Predictably, the helium skipped blithely past the metal interfaces in our equipment at all points tested. As a rare, costly, and highly unpredictable gas, whether found in the human body or a piece of functional hardware, helium deserves a respect that is rarely accorded. Perhaps this latest error will instill a new humility on the part of our designers and hardware merchants, and gaskets will be put only where they belong.

Although almost every error of design and fabrication that could be made had in fact emerged in the Sealab III habitat, we also were at fault for failing to impress on each of our contractors that they must design components that could cope with helium under very high pressures. Quality control was left to the vendors without supervision. The vendors did not do their job and we were consistently ignorant of the fact. Interestingly enough, however, certification of the Sealab III habitat would not present a problem. As the structure resembled nothing in the history of naval construction, the certification team had no safety guidelines. The diving system, however, involving the DDC and its controls, would pose a big question. By virtue of its complexity, and the obvious difficulty of training personnel, much more time would be required for safety certification and follow-up training.

October 6, I remembered, was the twentieth anniversary of the Valley Clinic and Hospital. Each year Marjorie and I had expressed in some fashion our good wishes and congratulations to the staff, and my mother had always attended the annual party as an honored guest. The hospital, which was built for less than $20,000 cash, continued to provide a vital service. My sense of values reeled when I considered that Sealab III, by comparison, would cost about $15 million.

A heartbreaker arrived for me at about this time: a letter addressed in feminine handwriting that made me cry unashamedly. It was from the widow of a crewman of the USS *Scorpion*, our latest submarine disaster, requesting my clinical opinion of the final minutes of her husband's life. The Navy had given her the correct and routine answer that her mate's death had been instantaneous and painless. She asked for an explanation of how this could be so. I could only hope that my answer would give her relief. Under the circumstances assumed for the losses of the *Thresher* and the *Scorpion*, death occurs within a matter of milliseconds at best, or seconds at worst. In the latter case, the individual becomes anesthetized so rapidly that there can be no sense of impending disaster. On the basis of my own experimental work, I was certain of these statements. Still, it was difficult to write these words to a widow, since we shun the thought of death in our fraternity. I can only pray that my response was helpful and perhaps will be passed on to other grieving families. Some days hurt like hell.

As in the space program, we must allow for the loss of human life. Whenever men and women are placed in a hazardous, isolated situation, the possibility of death exists. The principal investigator decides where to draw the line between acceptable risk and excessive hazard; the responsibility for the decision belongs to that person alone.

My Chronicle continued:

24 October 1968

After several weeks of frustrated delay, I finally departed the Knight's Rest in San Francisco today and settled in at the familiar Outrigger in Long Beach to await the arrival of our surface support ship, *IX 501*, from San Francisco. Up to this point, the delays have been a source of annoyance; now they pose a real problem. Human beings and equipment are both subject to the aging process; of the two, the latter is more vulnerable. We have been diving our hardware continuously for months and may well be close to the calculated life span of the gear. Long ago we abandoned hopes of a visit with our families on Thanksgiving, Christmas, or New Year's. Even so, morale is high. Whatever the aquanauts may lack, it certainly is not a sense of humor, and we need a hell of a lot of that every hour, on the hour. How much longer we can laugh off these delays I cannot predict.

31 October 1968

The presence of Jacques Cousteau has given us a respite from our nagging cares. He is the only man I have ever known whose charisma is indisputable. His group is here to film a 1-hour documentary on the Sealab III experiment. Cousteau and I agreed that the opening scene should be in a quiet, reflective mood, with the two of us exchanging reminiscences and philosophies about our undersea programs and about international efforts to draw attention to our deteriorating oceans. We selected a rock fireplace setting in the guest house on San Clemente Island, with an opening shot of the Sealab complex in its moor, dissolving to the two of us viewing the scene and smoking our pipes.

Later in the morning we shot introductory group scenes, completely obfuscated, it seemed to me, by welders' arcs, chippers' guns, and unprintable shouts from a nearby boatswain's mate. After several retakes, the impresario declared the filming a success. How the film crew could make anything out of that cacophony is beyond me, but I yield to their greater experience and to Cousteau's undisputed reputation as a filmmaker.

In light of our past experience, I am persuaded that most

Sealab III enters the water for a test off Hunter's Point Navy Shipyard, San Francisco. (U.S. Navy photo)

of our hardware contractors actually believe their own advertisements in the trade journals, which guarantee performance of helium unscramblers, thermal protective suits, diver communications devices, helium reclaim systems and, above all, underwater connectors. Even worse is the Department of Defense policy of forgiving and forgetting the performance failures of the same hardware contractors that have let us down time and again. Appropriate litigation is unheard of, and official censure of the vendors is not disseminated, as contracts are returned to the same offenders year after year. At best, we get free replacements of faulty components, but no compensation for the enormous cost of downtime when failure of a single component in our complex system inevitably causes a chain reaction in which other, blameless hardware items are damaged, to be replaced at our own expense. In all likelihood, the best solution is to award the entire contract package on a fixed-price basis to a single prime source.

12 November 1968

It is a good feeling to see our aquanauts reassembling in these last days of testing equipment at sea; we are all, however, restive. I do not work with them in the water for obvious reasons: I cannot match them in terms of physical conditioning or in the use of our new equipment. True, I can still do 100-foot breath-holding dives in a training tank, which most of them cannot match, and I hold the record for the fastest nonfatal free ascent. But these are stunts, carefully learned over years of work in the tank, which have no real relationship to diving capability. I like to believe that the friendship I share with the Man-in-the-Sea group, aquanauts and civilians alike, exists because I am a diver in my own right and one of the three who developed the system of saturation diving. For Bob Workman, Walt Mazzone, and me, this should be sufficient achievement. For almost fifteen years, I relished the role of big operator in diving and underwater escape. I chose to be the subject of my experiments, under the philosophy that the prime investigator should be the prime subject if adequately trained. But there comes a time when the hunter must return from the hill and the sailor ease back from the sea. Meanwhile, the Sealab program goes on. We are well into November; life does not stand still.

When I took on the job of moving into the quarters I share with Walt aboard the support ship, *IX 501*, I felt quite certain that beneath the pile of books, charts, manuals, and fishing gear must lie my bunk. Of one thing we were both sure: the two of us could not live peacefully unless some order was established. As a start, I agreed to keep my voluminous writings in a file ashore. Walt countered with one empty drawer under the bunk for my clothes, plus agreeing to hang his boots overhead. Also, he would install another desk and an overhead book rack. Casting about to top this largesse, I offered to take his dirty clothes back to the BOQ and launder them. (I was later to regret this, but at the time it seemed to solve the problem.)

28 November 1968

Valiant efforts have been made to ready PTC 2 for a certification dive, but all efforts until now have been thwarted by a series of electrical grounds and battery and cable floodouts. The NavShips certification team came and went, and finally today one member bravely returned to give us a last chance. The dive was to be made in the hydrostat mode, with an extra hatch installed to permit descent to the bottom with only 1 atmosphere of pressure in the vessel. After the occupants entered the PTC, some time was spent in securing the recalcitrant hatch. Much more time was used up in attaching the downhaul system and the 2,500-pound anchor. Finally the PTC hovered over the ship's well, ready to descend.

Snug in the jaws of the crane, she was lowered about 10 feet below the surface. At this point, the procedure called for the winch operator to take a strain on the SPCC (strength, power, communications cable), which would release the PTC from the firm grasp of the crane and permit the occupants to haul themselves down to the bottom under their own power. When the winch operator threw the lever into takeup position, the winch snarled and whined, but nothing else happened. The winch was out of commission. The crane lifted the PTC back on deck, the occupants disembarked, and a knot of worried men peered into the guts of the giant winch. Another dive had been aborted.

An hour later, the trouble was diagnosed: a pin had been sheared off the drive shaft. Lacking a spare, we had to ask San Clemente to machine another pin, to be picked up by the courier we sent. Conversation degenerated into grunts and expletives. Two hours later, the homemade pin was installed, the winch functioned, and the certification dive was made. We were happy with the results and a few lucky ones departed to spend Thanksgiving with their families tomorrow, 29 November.

30 November 1968

We were to lift the habitat off the barge today and place her in the water for tow alongside the *IX 501*. Camera crews came from all over the globe, even filling our airspace with helicopters. The giant floating crane was delayed in passage from San Diego but arrived to begin the lift about midafternoon. It was an impressive sight and highly photogenic. Inches at a time, the habitat was raised, while at each step our aquanauts swarmed over the complex, tightening the shackles, detaching hold-downs, and performing miracles of rigging with apparent disregard for life or limb. At last, the habitat was lowered to ride smoothly alongside the *IX 501*.

Not long after dark this evening, in the course of an unmanned, routine dive to the bottom, our only operational PTC experienced complete flooding with seawater at 590 feet of depth. As the winch began to raise the PTC, a few people noted that the cable showed considerable strain, but no one realized that instead of lifting a total weight of about 6,000 pounds, it was under tension of more than 30,000 pounds. The cable withstood the strain, and the PTC cleared the water, streaming large amounts of the Pacific Ocean from the sprung lower hatch. Once the chamber was opened, the full range of damage was evident. We could see at a glance that all electrical connectors and cables had been flooded or burned out by salt water; short circuits had occurred in the battery-powered system, and all instruments and communications gear were beyond repair. Complete repair and recertification of the ruined PTC will require sixteen weeks or, if we decide to jury-rig for limited function, we might be able to use the chamber in about a month.

1 December 1968

At a meeting of all hands and support personnel today, Jack Tomsky explained the situation and advised all but the aquanauts to return to their respective activities to await developments. There were few questions; the crew members understand the problems and hazards that attend such an undertaking, and they accept casualties without a whimper. After a representative from DSSP flies out to assess the situation, a decision will be made on the course of action to be taken. I opt for the short, jury-rigged approach, which would allow the aquanauts holiday leave, richly deserved.

Before I left the ship, I saw Walt briefly in our cabin. He is taking this one very hard, and small wonder. Of the entire crew, he has worked twice as hard as anyone over the past six months. Until last night's disaster, he could finally see the beginning of a successful operation—his last with the Man-in-the-Sea program. Now he must question whether he has enough ampere-hours left in his battery to take this. We have spent so many years working together that I could read his thoughts; talk was not needed. I left my old shipmate lying on his bunk, and picked up his laundry as I left the cabin.

Tonight the wind, which was fresh this morning, is blowing a near gale. All small craft have long since been secured, and the *IX 501* rides it out with a skeleton crew and the habitat plunging at her side like a frisky colt. At least we deserve better weather than this.

9

Sealab III
(February 1969)

Following our flooding casualty, the wise decision was made on 13 December to send the aquanauts back to their families for holiday leave and to proceed with the contractor work. I spent a blessed fortnight mingling with my family and grandchildren over Christmastide, which renewed us all.

The first week in February 1969, I joined Jacques Cousteau for an expedition party on board *Calypso*. I had some severe reservations—not to say guilt—about leaving the project, but when Captain Nicholson assured me that Bob Bornmann could cover for me during the ongoing certification procedures, I took the next flight to San Diego. I arrived there in time to say goodnight to Jacques, his wife, Simone, and their son Philippe.

By early morning on 1 February, we were out in the swells of the Pacific, which grew worse as the day wore on. I ate lunch with misgivings, not really feeling seasick but thinking about it more than was necessary. Immediately after lunch, I caught a glimpse of the chef making a dash to the port rail, and not a moment too soon. Strangely enough, the sight gave me a sense of improved well-being, since I was obviously better off than he. As the afternoon waned, I was further rewarded with the sight of three more seasick crew members, so that by the evening meal I felt positively radiant. Another guest joined us: Ted Walker, biologist, an authority on the gray whale and a student of nature in any form. Among the crew, I knew from previous encounters the oceanaut André Laban, and my roommate, Bernard, just returned from two months of dissipation in Paris.

On *Calypso* there were no distinctions between guests and crew; we all ate together, joked together, and lived with a complete lack of formality, which extended to the attire of guests and crew alike. Outfits were colorful, some ragged, all functional. Some preferred bare feet; some wore beards; all wore long hair save André Laban, who had a close-shaven pate. The sole crew member in formal attire was our chef, Marcel, who, befitting the dignity of his rank, always wore his high white hat and knotted white scarf, seasick or not. The cuisine on the ship was better and more imaginative than I have found in any French restaurant, bar none. Work was accomplished skillfully and without formal supervision. At supper Cousteau issued a written plan of the coming day's work, we discussed it, and everyone did his job the next day. As luck would have it, all were tea drinkers like me!

We approached the sheer cliffs of Guadeloupe Island, off Mexico's coast, by the early morning light of a setting full moon. From the foot of the cliffs, we could hear the barks, growls, roars, and explosive Bop! sounds of the bull elephant seals we had come to film for a documentary. The plan of the day called for a beach party of cameramen, grips, and other trade specialists to photograph the seals in their natural habitat. As a climax, a large bull elephant seal was to be captured, harnessed, and lured to the water, where he would tow Philippe through the waves while on his way to greater depths. The herd of young bulls chosen had apparently been isolated from the main colony because the bulls had developed the urge to breed. They were relatively small, about 1,000 pounds each, no match for the harem masters. They would live as a celibate splinter of society, practicing their fighting techniques, until one by one they would return to challenge the big herd bulls, perhaps by the next breeding season.

Just before noon, we began to slip our anchorage a bit, and shifted to a better spot to the south, immediately opposite a ravine containing the ruins of an old Mexican barracks. It sheltered a small group of itinerant Mexican lobster fishermen who promptly paid us a visit, presenting us with a crateful of lobsters and an outboard motor, described as "*Señor* Johnson *siente muy malo.*" The "*Señor* Johnson" was brought aboard and its carburetor immediately repaired as the crew's way of thanking the lobstermen for their generous contribution to our evening meal.

I caught up with the Cousteau beach party at about the moment when the bull chosen to star in the film was being secured to boulders with about 100 fathoms of nylon line and eight shots of chain. No drugs had been employed in his capture, and the seal appeared to be in good health. A harness was fitted to the body after ensuring that it would not impair respiration. The bull had numerous healed wounds covering his entire body, tokens of previous altercations with his fellow outcasts, and a recent, unhealed wound on the left flipper. Since the sun was now too low for photography, we left him in place until early morning.

The first film group departed for the captured bull at dawn but returned shortly with dejected faces. The animal was dead—killed by a massive attack of his contemporaries. In the many years of filming animal documentaries, this was the company's first death. They took it hard.

Characteristically, Cousteau determined to make the best of the situation. The goal of the elephant seal documentary was to present the life history of a single animal, from birth to its ultimate demise. As we now had a dead seal on our hands, we would stage the end of the documentary and film the seal in his final resting place, an underwater ledge of about 130 feet in depth, where a vortex of current had deposited a large number of elephant seal bones. We would transport the seal to the graveyard and start filming immediately. The burial scenes, filmed by torchlight, made a dramatic finale to the documentary.

Back again on San Clemente Island after spending several

glorious days on *Calypso,* I noted with a lighter heart that the schedule looked promising for a dropdown of mid-February. The certification dive of PTC 2 was completed in the manned hydrostat mode without a glitch. At last we had the official seal of approval, without which we could not have advanced one step further.

I had been invited to be interviewed by ABC TV, with the purpose of exploiting a recent magazine article in which Scott Carpenter had explained his medical disqualification for saturation diving. Caught off guard by a question referring to Scott's "malady," I stammered through the interview with what must have sounded like evasive action. I pointed out that we had carefully monitored bone X-rays of all our aquanauts, and that aseptic bone necrosis was notoriously a disease of caisson workers but nearly unknown among Navy divers or aquanauts. On other network news soon after, I was dismayed to hear that Scott had described this bone disorder as a common disease of Navy divers. He was undoubtedly misquoted, as we all frequently are, but we must try even harder to keep the facts straight.

Finally, by early evening of a rain-swept Saturday, 15 February, there remained only the task of lowering the habitat to the ocean floor. From my vantage point on the bridge, I could see Jack Tomsky below me, his hawk-like features drawn tight with the pressure of the task, flanked and rearguarded by Captains Nicholson and Ebel, while Scott Carpenter moved about like a defensive linebacker. From time to time, as the habitat inched lower into the sea, the winds brought to my ears shouts from the reporters on the sidelines, which came across as "Lower her away smartly, Jack!" or "Slack off and let her drop!" Of course, one does not lower such a mass rapidly, but the terse phrases must have had some effect, for Sealab abruptly sank from view and dropped 30 feet on her initial plunge until the counterweight took hold and steadied her at a safe level. Five minutes later, our little group huddled about the TV monitors listening to the cacophony of communications and awed by the beautifully clear view of the habitat going down smoothly in a transparent sea, her lights aglow. After nearly 3 hours, a hard, white ocean floor came into sight and the undersea home touched down gently, scarcely disturbing the sand. We howled with unrestrained delight.

Sunday morning, 16 February, dawned clear, but in the conference room on *IX 501*, I sensed that all was not well. During the night we had begun to lose pressure in the habitat. The numerous penetrations through the hull, although designed to withstand pressure differentials far in excess of the conditions of this experiment, had begun to fail one by one. Precious helium then bubbled out to the surrounding ocean. We were leaking gas at an insupportable rate, and corrective action had to be taken immediately.

Two choices were apparent: we could attempt to raise the 300-ton habitat, although preparations for this procedure would probably have to take place within the habitat itself; or we could send a team to the bottom to enter the habitat and plug the leaks so that the project could proceed. Admittedly, these choices meant a trade-off of maximum diver safety against a loss of millions of dollars of hardware, but, should the mission fail, the aquanauts could return to the safety of the personnel transfer capsule and a quick trip to the deck decompression chamber.

We sent *Deepstar*, a small two-man submarine, to investigate and pinpoint the multiple sources of gas leaks, so that the aquanauts who entered the habitat to make repairs could more readily locate the trouble spots. When this was accomplished, diving gear, wet suits, and detailed diagrams of the leaks were locked into the deck chamber with material for mending the leaks. The dive profile was set. Warrant Officer Bob Barth, who had dedicated almost a decade of his career to the Genesis and Sealab programs, was the previously selected leader of team 1. Berry Cannon, a civilian engineer from the Navy's Mine Defense Laboratory, a veteran aquanaut of Sealab II, and an authority on the habitat hardware, would perform the engineering feats necessary to survival of the habitat. The third aquanaut, John Reaves, our expert photographer, was also a veteran of Sealab II. The fourth member of the entry team was Dick Blackburn, a deep-sea diver with 14 years' experience and a qualified aquanaut.

At the same time that we began rapid pressurization of this four-man team in deck decompression chamber No. 2, a second team of divers was in the process of slow pressurization. The second group, accompanied by a physician, was destined for prolonged physiological studies at 450 and 600 feet in depth, prior to joining their fellow aquanauts in the habitat for the first phase of the project. In theory, we could move the physician via personnel transfer capsule from DDC 1 to DDC 2, if required. (In fact, a defective seal prevented this doctor's transfer when we urgently needed his services.)

The two teams began pressurization at the same time, although at different rates of compression. I had no concern about the rapid compression rate of our entry team, since I had personally compressed at the rate of a half-mile per minute on many occasions, without damage, while these men were descending at a rate of only 18 inches each minute. Having reached 620 feet of simulated depth in the deck decompression chamber, the divers were instructed to enter the personnel transfer capsule. When Bob Barth reported that he was "damned cold," we had the first hint that descent to this depth in the PTC could be physically detrimental. In the habitat, the devastating effects of this atmosphere are offset by high ambient temperature. In the PTC, however, it had been planned that the aquanauts would be protected by electrically heated suits. Only days before, I had learned that the suits presented an electrical hazard and could not be used. Since none of our water-heated suits were adapted for use inside the PTC, the entry team could rely only on partially compressed, helium-permeated, conventional wet suits for protection against the cold. Even this meager protection might have sufficed had all gone well.

The transit time of the PTC, from the moment of disengaging from the DDC to final positioning on the bottom, was about 2 hours instead of the 30-minute estimate I had been given. Worse still, some well-meaning person had removed the deflecting baffles from the carbon dioxide scrubbers, permitting a direct blast of frigid heliox on the aquanauts in the capsule. The result was that four badly chilled aquanauts reached the bottom near Sealab III. Barth and Cannon would go out on the first dive. Their job was to set and clamp the four leveling legs on the habitat, flood ballast tank 3, "blow" the skirt, undog and open the hatch to the diving room, enter, and repair leaks.

Within view of our remote TV eye, we saw Barth swim to the entrance ladder and open the valve to blow the skirt dry. Next, he appeared to hammer at the hold-down dogs to the hatch. Thinking that the dogs were stuck, someone topside increased pressure inside the habitat to help the loosening process. Bob tried to force the hatch open, but it was quite impossible. After a few minutes of fruitless effort, he became aware of Berry's absence and went back to the PTC, where Berry had returned a bit before. Back on deck, the aquanauts transferred to the DDC and drank hot fluids, showered, and covered themselves with electric blankets. Bob Barth responded to my anxious query as to how he felt: "I was goddamned cold and I'm sorry I didn't finish the job!" The time was 2100 hours on 16 February.

I went topside to report to operations that our aquanauts were terribly cold and their core (inner body) temperatures must have dropped to a disturbing level. Time would be required to restore this deep-seated heat loss. On the other hand, we were fast running out of gas, as we had already used nearly a half-million cubic feet of it. A relief barge carrying a puny 100,000 cubic feet would not arrive alongside for hours. The race against time was clearly not in our favor.

We considered using the fresh diving team from DDC 1, which had by then compressed to 600 feet. From a logistics and safety point of view, however, the switch seemed unwise and was dismissed. I was now asked the obvious question: How soon could we dare to expose the original entry team to the thermal stress of another dive? I could have cried at the lack of basic research that, had it been authorized years ago, would have provided the vital data. I could only hazard a guess that they could make the dive after 4 hours without too much risk if the safety checkoff procedure was shortened and hot water–heated suits could be rigged up for the divers to wear in the PTC.

I must add that Scott Carpenter asked for more time before the second dive, in view of the expected arrival of the small gas barge. We overruled him, however, and at 0315 hours on 17 February we sent the team down again. Once more, the four men entered the PTC, went through the lengthy checkout procedures, and sat shivering in their nonfunctioning water-heated suits as the capsule was dropped about 60 feet from the habitat. In the range engineer's cabin, I watched the TV monitor and listened to communications. Almost 5 minutes passed, yet no divers had come into view. A web of fear began to close in on me and I fought it back. We have done this so many times at this depth at the Experimental Diving Unit, I told myself. But this wasn't the EDU, and my divers were late. When I saw Bob Barth swim to the ladder, I felt better. Alone, he climbed up and pushed the hatch, but it would not yield. Not wasting time, he swam off camera to get a crowbar. There was still no sign of Berry, and the fear came back, stronger this time.

I said, very loudly, "Goddamn it! Where is Berry?" No one answered.

As Bob returned with the tool, I saw him look to the opposite side of the habitat, drop the crowbar, and swim furiously off camera. Within seconds we saw a tremendous boil of silt on the TV screen, and now fear became an agony of certainty. No trained diver moves so fast as to stir up a blinding boil of silt unless there is real trouble. In a second, Bob returned with Berry in his arms. He tried to hold Berry's head up in the gas pocket of the skirt, but that didn't work. He attempted to force the buddy-breather mouthpiece into Berry's mouth but failed on three tries, stabbing against teeth locked in a final convulsion. Bob then turned and began to drag Berry to the safety of the PTC. I rushed below to the main control console, from where, miraculously, the correct orders to prepare for the crisis had already gone to the PTC. Dick Blackburn was garbed in time to help get Berry inside. Bob Barth had made it inside and was safe. John Reaves and Blackburn practiced turnabout on mouth-to-mouth resuscitation and external cardiac massage, buttoning up the PTC for emergency ascent, and communicating topside. We started a rapid emergency liftoff once the lower hatch was confirmed sealed. But 14 minutes later came the unbelievable message from the PTC: "Berry Cannon is dead."

Regardless, efforts at resuscitation continued. Ninety minutes later, the seal with the DDC was made, and three of my four aquanauts returned to warmth and life. It was daybreak on 17 February, or maybe full sunrise, but nobody cared much anymore about the time. The stunned and exhausted aquanauts in DDC 2 followed the instructions they were given for use of a burial bag brought from the island, finally placing the bag with its contents in the outer lock for solitary return to the surface. There, the four diving rigs were detached from the PTC under guard and placed under lock and key. Bill Liebold began a methodical collection of video- and audiotapes, logs, and notes to be impounded for the investigation.

Hours later, in San Diego, I assisted the pathologist and toxicologist in an autopsy. We found no real clues to the cause of Berry's death. It would be several days before blood determinations, microscopic examinations, and toxicological results would be available. I returned to Point Loma, checked into a motel and, with a heavy heart and little to say, called Mary Louise Cannon, Berry's wife.

The next day, 18 February, I was advised that a new development had occurred in the tragedy. I was to return to San Clemente and talk with Comdr. Bob Bornmann upon arrival. Unfortunately, seas were running and small-boat trips had been suspended. The new information could not be passed by telephone, so I would have to wait until morning.

On 19 February tragedy piled upon tragedy. My God! Was there no end in sight? In response to my urgent plea to examine the CO_2 absorbent in all four of the Mark IX canisters, Bob Bornmann and Bob Thompson had unscrewed the caps on each one. The canister on the MK IX rig that was known as No. 5, presumably worn by Berry Cannon, was found to be empty! At 600 feet, a human wearing that rig could not remain conscious for more than a couple of minutes if the exhaled CO_2 was not absorbed, then would convulse and die shortly afterward. That was the exact scenario of Berry Cannon's death.

We called Captain Nicholson, informing him of Bornmann's report, which as yet we had not personally verified. Unbelieving at first, he was at last partially convinced and ordered Bill Liebold, Dick Garrahan, and me to personally examine the four MK IXs involved in the fatal dive, make careful notes, and call back our findings. That could not be done at once, as Dick Garrahan was attending the funeral services in San Diego. Emergency calls were made to arrange travel to Washington, D.C., in company with the four rigs, following our preliminary inspection. Meanwhile, Jack Tomsky and I had to face a previously scheduled interview with *Time* and *Newsweek* magazines. I had just enough time to call the toxicologist in San Diego, informing him that we were particularly interested in possible CO_2 poisoning. He told me that blood specimens had gone to the university laboratories for special analysis, with strong emphasis on CO_2. Perhaps the results could be obtained by the next day, 20 February.

When Jack and I met the reporters 30 minutes later, the question arose concerning the possibility of gear failure. Jack's answer was basically true. He stated that upon arrival at the surface, each MK IX rig had been tested and found to be functioning properly—delivering gas flow at an apparently normal rate. The rigs had then been impounded for shipment to Washington, D.C., for more extensive examination and testing by unbiased experts. Since a formal Board of Investigation was being convened, it was normal procedure to impound all items of pertinence to the case, including tape recordings, log books, videotapes, and the like. Jack further stated that, to his own knowledge, the diving gear had been in proper functional order at the time of the accident, but he also acknowledged that equipment failure can and does occur. For my part, I went over the results of the autopsy and pointed out that the medical investigation was less than half complete, with at least five days before all the answers were in.

Ten minutes later, Bill Liebold, Dick Garrahan, and I were in a small room in the administrative building ready to inspect the suspect diving gear. We debated quietly whether we should have Chief Paul Wells perform the breakdown and checkout, since we all knew that Paul had major responsibility for all aquanaut diving gear and would be a "party of interest" in the forthcoming investigations. We called Paul in and asked him to conduct the outlined checkout procedure as we took notes. The basic features to be tested were bag flooding, gas flow rates, bottle pressures, buddy breathing accessories, bypass flows, and canister inspection.

The first rig was examined and found to be in good order, save for wet CO_2 absorbent, probably incurred during the ascent, when the rig had been strapped externally to the PTC. The second rig, No. 5, passed all preliminary checkouts. Then Paul removed the canister that was to hold the supply of CO_2 absorbent.

"This is an empty canister," he said as he raised it clear of the pack. Unscrewing the top, he peered inside and offered it to me for inspection. It was indeed empty, although I could see clearly in its interior the outlines of Berry Cannon's death certificate, which would no longer read "Cause of death unknown."

Paul stated flatly that he alone had been responsible for filling this and the other canisters with CO_2 absorbent. No excuses, no evasion. Not one of us believed that Paul Wells would ever load an empty canister into a MK IX rig. The inspections proceeded, but they were anticlimactic. The rigs were locked in individual boxes for air transport, and we gathered our notes for another call to Captain Nicholson.

In Washington, the decision was made to release our findings within two days. We protested that this would demoralize our group and jeopardize the safe completion of the long decompression of the eight aquanauts remaining in the deck chamber. But the decision on the timing was final, and we must be completely aboveboard with the press. Our last chore was to call the *Time* and *Newsweek* reporters, telling them to hold their stories pending a new development.

What a day and night, so close upon other sad days and nights. To a man, we were certain that Paul Wells had made no mistake when he loaded and certified the diving gear. Unnamed persons had helped Paul prepare the Mark IX rigs on Valentine's Day. One of the outfits had not been loaded by Paul, and I could only conclude that that rig was the one with the empty canister. Just who the "helper" was may forever remain a matter of speculation. Paul did not remember, and no one volunteered to fill this void in the scenario. Being the man he is, Paul pressed the matter no further. Simply stated, the rigs were his responsibility, and he would not divert the blame.

10

The Ocean Simulation Facility
(1970–1972)

Moments of truth do not always come in blinding flashes. More often, truth is acquired through gradual, sometimes painful steps. The momentum of the Sealab projects diminished as consensus indicated that, in the eyes of the Navy, the mission had been accomplished. The ancient dream of humans living on the ocean bottom, free to move and work for hours instead of minutes, was no longer a whimsy of science fiction. The supposedly immutable barriers of physics and physiology that had shackled divers over the centuries had at last been hurdled as a direct result of our laboratory experiments, and then conclusively proven in the open sea. As the lengthy Board of Inquiry investigation into the cause of Berry Cannon's death slowly wound down, so did the Sealab project.

I left directly for Great Lameshur Bay off the coast of St. John Island in the Virgin Islands, happily in the company of my wife. There I was to consult on Tektite I, a project designed to put four scientists for sixty days in an undersea habitat furnished by General Electric. The habitat was anchored on the ocean floor 40 feet below the surface in warm, clear, rich water. The four scientist–aquanauts, Rick Waller, Conrad Mahnken, H. Edward Clifton, and John Van Derwalker, followed a program set by the Department of the Interior, taking a census of species of organisms in the area, observing the behavior patterns of fish, and detailing the habits and migration of the spiny lobster, the source of that popular seafood known as lobster tail. NASA conducted behavioral experiments to prepare for future journeys into space.

Sponsored by the U.S. Navy, which provided surface support, overall management, and my services as saturation diving consultant, this joining of science and the military to explore ways to exploit the fruits of the sea resulted in a monumental amount of data regarding fish behavior and the geology of the ocean floor. Over a quarter-million facts and clues were recorded relating to the aquanauts' mental alertness, muscular coordination, and social behavior. After 60 days' submergence, the aquanauts left the habitat for a 19-hour decompression in a chamber on the support barge. By any standard, it was a completely successful, rewarding project to the several sponsors, to all participants, and even to President Richard M. Nixon, who cabled his congratulations on this "milestone in human achievement."

I enthusiastically anticipated my next assignment: contracting the software for the Ocean Simulation Facility (OSF), a new deep-diving tank at Panama City, Florida. Life in Florida was incredibly good. Marjorie and I rented a perfect house right on the water and spacious beyond belief after a lifetime spent in motels and shared quarters aboard ship. We had a boat and unlimited vistas. An added blessing showed up in the form of a 4-foot pet alligator we named George Foote III, who lived nearby, ate hot dogs, and came at my call to astound my visiting grandchildren.

The planned hyperbaric chamber was part of the Navy's move to consolidate deep-diving training and experimentation at the Naval Ship Research and Development Laboratory (NSRDL). It was an agreeable opportunity to use my experience of many years with deep-dive tanks and saturation diving decompression tables. Incidentally, it also extended my service in the Navy, as senior consultant on undersea habitat existence. I chronicled special events in correspondence addressed to one or another of my close friends and sent out copies to a growing list of old cronies, new acquaintances, professional allies, and medical colleagues, all of whom I wanted to keep in touch with on a personal basis. As it turned out, this was a natural extension of my long habit of journal-keeping and served to express a great deal of personal philosophy during my final years in the service of the Navy.

I have never shirked the opportunity to voice my personal reflections on a given subject. My response to the request of Lt. Comdr. Victor C. Evans, USN, Office of Chief of Naval Operations, for my opinion on potential uses of the Sealab III habitat was that the habitat had not been conceived as a weapons system and would not be considered as such in my evaluation, although it certainly had the potential to operate as one. As I wrote to him:

> It is first and foremost a useful tool to gather large amounts of oceanographic data of value to naval missions and scientific programs alike. At the present time, our Navy is shortsightedly thinking of application of saturation diving concepts solely in terms of salvage. This points up a singular lack of imagination. Even within this narrow frame of reference, the bottom-placed habitat can do a better job than the MK-2 or MK-1 deep-diving systems, if we think in terms of man-sized salvage jobs. Surely, the capacity to put a salvage team of a dozen or more divers immediately adjacent to the sunken hull, with plenty of room for equipment, to work in teams of two or more around the clock—in disregard of weather—are advantages that cannot be denied. But the seabed habitat is most important as a tool for scientific investigation of phenomena relating to the sea floor of the continental shelves, the physical characteristics of the surrounding water, and the animal and plant population of this marine environment.
>
> The in situ use of a habitat will assuredly find direct ap-

plications in the fields of marine biology, marine geology, ocean climatology, seabed mining, and undersea construction, to mention only a few of the many disciplines that could be well served by such structures. Permanent installation of single or multiple habitats on selected shallow sea mounts and a few sill areas of the world's oceans could, over a period of years, provide oceanographic data that cannot currently be matched. Additionally, these emplacements could serve as continuous monitors of undersea traffic, and in some cases could be a vital component of undersea ordnance test programs.

It may readily be said that further work with the seafloor habitat should be the sole province of civilian–scientific disciplines, but this is a dangerous bit of rationalization. Clearly, the talent to build and operate undersea habitats will be the Navy's province for some years to come. Today, our Navy is disinclined to pursue further habitat development and work. This is unfortunate, as at least ten major foreign nations are proceeding with habitat programs, all utilizing the basic data supplied from our efforts and made freely available to them.

I expressed my opinions to the higher-ups in ONR in this plain fashion and could go no further except to rant at them all for burying their heads in the sand if they did not follow this course of action.

But at the NSRDL, thanks to dedicated individuals both military and civilian, programs of imagination had evolved. In cooperation with the University System of the State of Florida, we initiated the first class of civilian–scientist graduate students of both sexes by way of an accredited university course, "Scientist in the Sea" (I called it—with Teutonic overtones—the SITZ bath). Although all ten were certified scuba divers, Bob Barth quickly put this obstacle aside by training the students to be trustworthy undersea workers in any gear. All ten had earned undergraduate degrees in scientific and engineering disciplines. Our students were exposed to the broad panorama of training in techniques applicable to each one's specialty, offering twelve graduate credit hours toward Ph.D. degrees in marine biology, ocean engineering, or oceanography. Our twenty-eight lecturers for the didactic portion of the course were virtually all at the M.D. or Ph.D. level, and visiting dignitaries came for evening seminars, usually at their own expense.

To me, "Scientist in the Sea" was one of the most exciting breakthroughs I had seen in years, and would surely lead the way to a healthy coordination between the military and academic communities. It even had the blessing of higher echelons in the Navy, though none save Capt. Bill Nicholson was willing to commit himself to the written word. To me, this attitude was old hat: if we fell flat, higher authority could chastise us freely; if we brought it off, they shared the achievement. I was grateful for the presence of Captain Nicholson, who took the career risk to back his convictions.

Toward the end of 1970 I wrote a long letter to Rear Adm. Ralph E. Faucett, Bureau of Medicine and Surgery (BuMed), to review the role of the medical officer in the Navy's diving program and to offer some thoughts on advanced program planning for officers who elect this way of life. I reviewed the work of the pioneer Navy physicians in the hyperbaric field (French, Behnke, Yarborough, Van der Aue, and others), who in 1935 were concerned primarily with matters of diving physiology, and to a lesser extent with submarine medicine. Their area of concentration was understandable, since submarine personnel rescue involving deep-sea diving was a critical priority. Also, they had embarked on an ambitious program of helium–oxygen diving that presented a new set of problems, best solved by a good medical-engineering approach.

The era of the diving doctor had lasted about two decades, ending around 1955, when emphasis shifted entirely to submarine medicine, in response to our new nuclear-powered submarine fleet. Diving medicine was then very nearly phased out of the picture. Although I would be the last to deny the need for medical input into the field of nuclear submarine habitability, we overreacted in establishing two medical billets on board each Polaris submarine. (In 1959, I was called to BuMed and informed that henceforth there would be no diving medical officers in the Navy.) Perhaps our overreaction prior to 1958 was understandable, since until that year the possibility of effective diving activities below 200 feet was inconceivable. With the advent of saturation diving, there came a radical reversal of this philosophy.

I rested my case with the following prediction: "Today we can dive to 1,500 feet and I am convinced that within a few years even greater depths can be reached. For safe and orderly progress into these depths, we will need a sensibly large cadre of well-trained diving medical officers and paramedical personnel assigned to carry on the required work in basic research with adequate support and without threat of dislocation because of a lack of billets."

I pointed out to Admiral Faucett that by 1964 only three Navy doctors had been trained in saturation diving: Workman, Bornmann, and I. We had since trained about a dozen more, but there were nowhere near enough billets for diving medical officers (DMOs) in operational and basic research areas. I anticipated a reduction in our future needs for submarine medical officers (SMOs) and urged that a transition be made to provide at least a fair number of these billets for DMOs.

After I reviewed in my letter the truly incredible advances we had seen in diving sciences and engineering in the past decade, I suggested that we needed to fill the ranks of investigators who would work hand in glove with the new generation of undersea engineers. Furthermore, with the establishment of the National Oceanographic and Atmospheric

Administration (NOAA), the Navy in my opinion had a clear dictum to take the lead in undersea research, working jointly with industry and the academic community. My letter then expressed my hope that Admiral Faucett would give these thoughts gravest consideration and become convinced that the time was ripe to adopt some of the changes I had proposed.

When I received a heartening response from Admiral Faucett to my prodding letter, I followed up with a letter to Rear Adm. Maurice Rindskopf, Office of Naval Operations, to offer both my assessment of the current status of diving biomedicine and some proposals that the Navy might adopt. I summarized my pitch in one paragraph:

> Looking realistically at research and development in deep-sea diving, we should honestly examine our own deficiencies over the past decade. Our problems of decompression revolve entirely around physiological knowledge of gas transport in the tissues of the human body. To this day, we have never really obtained that knowledge, largely due to our lack of tissue-gas measuring devices. In all probability, this barrier will be passable within five years. With moderate money and high priority, it is quite possible to identify the specific gas-exchange characteristics of many body tissues. From these data, we can wisely computerize saturation and subsaturation diving decompression, and vastly improve our facility in deep subsaturation dives down to perhaps 600 feet. But it will take both dollars and time, and the latter keeps slipping away.

I wrote to Admiral Rindskopf in this bitterly frank way because he was an old shipmate. I had become increasingly aggressive about the Navy's neglect of the saturation habitat programs, and I had acquired a degree of immunity to repercussions since taking my name off the selection list as a guarantee that I could stay on this job until we put the first man down in our new chamber. I soon learned that my billet, destined for transfer, had ironically been sacrificed by someone in DSSP. In consequence, and in near parallel with Nathan Hale, I found myself outside the Navy, yet still on the payroll.

In the spring of 1971, we awarded the last contract for completion of the Ocean Simulation Facility, which is not likely to be duplicated elsewhere in my lifetime. The building to house the complex is one of the finest examples of military construction in this country. The massive chamber complex, three stories high and so large it staggers the imagination, would be delivered in May. The next task for me had to do with preparation and review of the tons of software vital to safe manned operation of this awesome system. I loved every minute of it.

As I was committed to a long "Scientist in the Sea" seminar at the University of Florida at Gainesville, I did not plan to attend the meeting of the Undersea Medical Society to be held in Houston in May 1971. Bob Workman, however, exerted constant verbal and written pressure to the point where I agreed at least to catch the excellent papers to be presented. I almost seized the chance to leave before the scheduled luncheon, but Heinz Schreiner persuaded me to stay and say grace. That was irresistible bait, and I swallowed it whole.

I had uttered appropriate words of invocation and settled down to nibble a few bites when Walt Mazzone arose from his seat at the head table and began a discourse that was most flattering to me. I finally realized that I was being presented the Albert R. Behnke Award! I was speechless. To say that I was honored by this award for contributing to advances in the hyperbaric biomedical field would be a gross understatement. Simply to receive professional mention in the same paragraph with the name of Al Behnke, the Navy physician who pioneered work in hyperbaric medicine, would be honor enough; to receive the check for $1,000 was frosting on the cake.

Returning to Panama City in a state of euphoria, I fed my pet alligator his dose of frankfurters, fertilized my tomato bushes, and took off for Bat Cave, North Carolina. The mountains were beautiful as always, but I was subdued and saddened by the sight of my cabin, which had been virtually destroyed by a group of vandals. With the old-fashioned help of some friends and neighbors, we rebuilt the cabin on Bear Branch (save for the privy, which must be relocated). By the time I had done all of this, my leave was over. I drove home to Alligator Point, deep in contemplation of the strange twists of fate we experience as we walk through life. If I had been forewarned that my cabin in the mountains would be vandalized, I would have denied the very possibility.

This disappointment, however, was trivial in light of my mounting concern about the status of our bone necrosis survey program. There was a possibility that our system of routine transfer of naval medical officers would cause us to lose the continuity of this project. The responsibility for overview of the potentially crippling disease had shifted from one hand to another, coming to rest in the capable headquarters of the Armed Forces Institute of Pathology (AFIP). At designated hospitals, radiographs of our aquanauts were taken and sent to the commanding officer of the Submarine Medical Center (SubMedCen), Groton, Connecticut. That office in turn routed the X-rays to three specialists in the United States and abroad for intensive review. The report then went into our files for collection by the AFIP. For a two-year period, we had attempted to contact all saturation-trained divers in the Navy, retired or otherwise, to build a complete record of bone surveys, starting with the Genesis series. We were missing data only on Bob Sheats, Andy Anderson, and Scott Carpenter, all of whom had promised to fill in the gaps. I was desperate to clarify the matter, as the question of bone necrosis in saturation diving weighed heavily on my mind.

Although the old workhorse Sealab I now rested on the bottom offshore Panama City, the fate of the Sealab III habitat (to be renamed NOAA I) still hung in the balance. The promise was that she would not be scrapped or disposed of in an ocean grave, but transported to the NSRDL for our eventual use. Of course, we had no money for this project, but I intended to make it a program of national interest and beg the money from civilian sources. The two habitats would then be operational, capable of extensive use for research and training, even though higher authority wanted to deep-six both houses. To me, that was a sorry way to bury mistakes, and I said as much. Besides their practical worth, the habitats were mighty big objects to be swept under the rug.

During September 1971, in my annual effort to requalify as the most ancient diver in naval history, I ran into a plethora of snags. My SMO, Jerry Weaver, wisely decided to do an "executive profile" blood chemistry test on me, and the results were less than rewarding: gout, and quite possibly a touch of prostatic infection (not gonorrhea), not to mention a hint of latent diabetes. Of course, I discounted all of this as idle laboratory rumor, but I thought BuMed would be a bit more uptight. In fact, I feared I might have to continue diving activities in my spare time, without official sanction. I prayed this would not be the case, since achieving a twenty-year career of diving would have meant a lot to me. In any event, I had my own open-circuit gear and planned to continue to enjoy nefarious undersea ventures.

Given the overall health situation in our family at that time, I predicted an inevitable loss. My mother-in-law, Dammie, who had lived with us for many years, was on the downhill skids with cancer, although without significant pain. She probably had less than a month to live. Her hospital care was excellent, however, and she would be spared any final agony. Meanwhile, Marjorie and I watched and waited, and I curtailed all travel.

Toward the end of August, I had discovered a painful tumor in my right breast. Thinking this to be the result of wearing scuba gear, I brushed the matter aside. The condition persisted, however, and the swelling enlarged almost to the point of embarrassment. I sought advice from our local Air Force surgeon, thinking all the while that a radical mastectomy would surely be in order. My illusions were dispelled when he declared that I was simply undergoing a moderate hormonal disturbance that would pass in due time. I hoped that with a restoration of my normal flow of androgens in "due time," the situation would reverse and I could burn my bra and join the women's liberation movement.

The first element (wet pot) of the high-pressure complex arrived by barge about that time and was installed in the Ocean Simulation Facility. The second component of this structural hardware was the most awesome combination of steel that I had ever witnessed. Though the whole thing struck fear in my soul, I gloried in the fact that it would be the most advanced and useful hyperbaric facility available for decades to come. More important, it would be available to all those in qualified areas of national interest and to investigators from foreign nations as well.

Dammie Barrineau continued to hang in there, and Marjorie and I maintained a dedicated watch. As sometimes happens with terminal cancer, she was past the point of physical pain and had now reached the last plateau of physical existence. Her mind was intermittently clear, and she was quite aware of her situation though she did not comment on it. I was impressed that she exhibited not so much a will to live, as a rejection of death. At the end, which was merciful, Marjorie and I shared the need to remember her well and then adopt a positive attitude and go on to other things.

Looking to happier matters, our daughter Judy was well established in her second pregnancy by the end of 1971. As our other five grandchildren were female, Judy planned to name her next, if it was a he, Adam. My alligator, George III, provided a certain distraction and grew apace, thanks in part to my daily feeding. He was shy with strangers, but finally accepted the presence of Bob Barth. Until that point, Bob had denied the very existence of my pet reptile, asserting vigorously that the whole scenario amounted to no more than a drunken medical officer throwing hot dogs to a floating log. Even that skeptic finally learned better.

A month into the new year of 1972, with fanfare and impressive show of brass, we were converted from NSRDL to simply Naval Coastal Systems Laboratory (NCSL), Panama City Beach, Florida, with a broader mission in respect to diving, salvage, and riverine military missions. The changeover called for ceremony, and that we got. We met in the heliport hangar, with about four hundred seated and two hundred standing, all chilled to the bone and moderately disgruntled. By rare misfortune, the weather had turned really bad for the first time in the nineteen months I had been there. Inside the hangar, which was unheated, lips were blue and sabers rattled. As I had just come from monitoring one of the many unsuccessful dives in a catalytic-warmed diving suit, I had a chilly start on the other guests. Thanks to the wisdom of our commanding officer, I was assigned to a seat rather than standing room. The formal changeover ceremony was impressive and touched the deep chords of pride and patriotism in all our hearts. As I was not sure of protocol regarding my hat, I developed the nervous reaction of removing it from time to time when I thought it appropriate. The gesture must have been infectious, as others around me began to do the same. I finally nested it under my armpit and resolved to upgrade my knowledge on such ways and manners.

Along with the new name for the laboratory came a new technical director, Jerry Gould, with an extensive background in undersea environments. A few years downstream, we would undoubtedly need a hyperbaric medical director, which our budget did not permit at that time. I thought I might try for this job once I became a civilian, as my spinal cord dictated either surgery or retirement from active diving.

To my chagrin, I was informed that Sealab III would be towed to the Santa Barbara Channel to be scuttled in extremely deep water. I will not recount the stupid rationalizations put forth to defend this decision; the plain fact was that the most sophisticated—and yet untested—undersea habitat of our time was scheduled for oblivion. Although I had asked for salvage of the hardware, the final word revealed that the hulk had been plundered. Piping had been cut clear with power saws, the ports awarded as souvenirs—not even to Sealab personnel, but to curio collectors. Certain instrumentation was purloined by programs that could not possibly use it, as it had been designed uniquely for the habitat. Neither written nor spoken promises were honored, and the whole affair sickened me.

But the old Sealab I had rested trim and shipshape on the ocean floor for one full year and we used it in the SITS program in the summer of 1972, with support from the Sea Grant treasury, the Naval Coastal Systems Laboratory and the University System of Florida. The second "Scientist in the Sea" program embraced sixteen students, including four women, for twelve weeks of training. We expected the program to establish a continuous pollution study to be monitored and documented by the graduate students working with our Navy mule-haulers and civilian scientists. To a person like me, long despairing of a relationship between Navy expertise and national interest programs, SITS was the first step toward an exciting coventure. There were at the time a host of government agencies with basically similar ecological aims, but of such disparate personalities and philosophies that mutual cooperation was not sought and was even shunned.

I had become increasingly distressed to see the dichotomies that split these groups, who could not comprehend the certain fact that every identifiable pollutant, wherever it is deposited, will ultimately come to reside in the oceans of this world. First will come contamination of our estuaries, with destruction of spawning grounds of our future animal protein sources. Should the human race survive this horror, our poisonous effluents will ultimately affect that thin layer of barely visible, open-sea phytoplankton (microscopic plants) that provides nearly 70 percent of the oxygen we need to live. I was one with my friend Jacques Cousteau, who keeps his caretaker's eye on the oceans of the world and gently chides us to safeguard the infinite future of this heritage. Especially did I feel this at the birth of our sixth grandchild, Adam.

By September 1972, a continuous but tolerable pain in my lower back had developed into an intolerable disability and I again submitted to the surgeon's knife. Despite the unwarranted optimism I expressed to family and friends, all did not go well with me. Almost immediately after I returned home from Tyndall Air Force Base Hospital, stitches out and ready for an indulgent convalescent leave, my right groin swelled to an inordinate degree. I tolerated the abdominal bulge and elevated temperature until the weekend ended, then presented myself to Cleve Thompson, M.D., my surgeon, on a Monday afternoon late in September. Of course, I sought effusively to convince him that what I had was simply a massive hematoma, but Cleve immediately demonstrated that, first, he tells the truth and, second, he is a good surgeon. He said the problem was a postoperative infection, and a few strokes of the knife proved his point with an issue of about a half-pint of very unpleasant material that thoroughly contaminated the examining room. With considerable aplomb, Cleve probed deeper, and took a culture sample of the second unpleasant geyser. On the next two days I returned for repuncture and drain replacement. On Wednesday, Cleve incised a lower portion of the wound, with copious reward. Next day I returned, draining like a stuck pig, and Cleve appeared very concerned. It seemed that the original culture had not been checked for anaerobic bacteria. When the aerobic cultures proved negative, the laboratory belatedly recultured under anaerobic conditions, and came up with gram-negative rods and spores. Lacking gas in the tissues, and no serious necrosis of tissue, the vote went for *Clostridium tetani.*

Cleve made a fast decision to open me up again—within the hour and without recourse to general anaesthesia. Since I had neglected my tetanus booster for nearly four years, I got a shot of that, then 30 million units of penicillin intravenously over the next 24 hours. The broth culture was sent to several laboratories, and I lay in bed with wound agape. Some days later, Cleve and I agreed on a short course (two days) of hyperbaric oxygen therapy in our Naval Coastal Systems Laboratory chamber. Word went about the diving locker that this was a cheap ruse on my part to justify my chamber pay for the month—a base and vile lie, as no amount of chamber pay would have been worth the experience.

I survived the treatment with only minimal attacks of lockjaw. I returned home once more, with a deep granulating wound in the right groin, and awaited the final bacteriological and toxicological test results that would declare me saved. At least I was no longer in isolation, and no one was sent to walk before me ringing a bell and crying "Unclean!" at each step.

11

"Scientist in the Sea"
(1973)

In many ways, the summer months of 1973 were both some of the best and some of the worst I have ever spent. Our garden in Panama City, thrice flooded, was slowly revitalized. We were now sole proprietors of the least fertile quarter hectare of land known to horticulturists. Once I accepted this fact, I then found that daily physical labor in the garden tightened the lax gut and gave illusions of youthful vitality. I walked more easily than I had for five years, and I was thankful for the success of the 1972 surgery. I was also able to sleep well. Admittedly, my surefire 80 percent disability pay was rapidly going by the board, but Marjorie and I preferred health to pelf. The joy of being able to walk two or more miles a day transcended my festering desire for revenge on the Internal Revenue Service. Live and let live, I said.

My spirits soared after a truly delightful ten days in June in the Bahamas (authorization orders—no cost to U.S. taxpayers), on my annual junket to lecture and socialize with some thirty-odd scuba diving physicians from all over the United States. The 1973 version was incredibly good. The hotel was superb, Marjorie was with me, we knew all the good eating places, the class of physicians was the best ever, the weather and water were equally perfect, and we finally got the hang of the island bus service. More important, the native Bahamians had become warm, friendly, courteous, and likable! When first I heard the phrases "You're welcome" and "Thank you" from them, I sought the reason for the incredible change of attitude in a population so long practiced in the art of insolence and studied neglect of mainland visitors. Somehow I suspected that M. Pindling, the elected ruling honcho, might have passed word that, within sight of their new upcoming Independence Day in July and the subsequent loss of British pound support, the islanders might treat tourists better.

Once again I returned to Bat Cave for a summer vacation. I have found over the years that a return to my mountain home is more than a happy event; it puts me back into an amenable perspective. My old friends and patients remember me as "Doc George," and it was a wonderful experience to move about the territory that I had once covered, and to revisit old friends on the ridges and in the hollows. The day of the frontier doctor in our country is probably near dead, and I cannot truly grieve its passing. But I never regretted a day of my life as a country doctor, and I always delighted in my annual return to Bat Cave.

I would spend a while visiting with old friends and former patients and walking the deck of my beloved Valley Clinic and Hospital. Sooner or later, however, I would go to my little cabin on Rocky Broad River, overlooking a beautiful waterfall and a deep pool. Three decades before, I had joined two great friends to build the cabin on that bluff, and the three of us gathered there regularly to spend the night, sing songs, and talk of life. Those two, Fate Heydock and Lonnie Hill, were gone—one by leukemia, the other by his own hand—and the cabin was a rebuilt version, but I returned when I could, to find coherence in disparate thoughts.

When I visited the cabin in 1973, I found it once again gutted and desecrated. How could this be? The carnage was very likely accomplished by young people whom I had delivered into this world, and their actions covered up by the very folks I had treated over the years in my practice as a country doctor. In all fairness, I could not blame either generation. The young ones somehow needed to destroy anything standing, and the parents wanted to avoid disgrace. It was only a rustic cabin on the bluff over a waterfall where, as a very young man, I camped in awe of the surroundings. Perhaps I would do so again. Whatever motivates young (or even older) people to do such acts of destruction is beyond comprehension. No matter, I told myself, the cabin was just a shelter and would be rebuilt; the bluff and waterfall would withstand all things, even time.

Back home, I had a day of lectures to the senior class at Florida Atlantic University, a real winner in my eyes. This academic institution is dedicated to producing first-rate ocean engineers, and was doing the job to perfection. Despite popular pessimism about the future of our undersea scientist–engineers, I was pleased to hear that these graduates were in demand in the marketplace. Although beginners' salaries were not high, at least a professional life in the ocean sciences was possible, which was not the case a few years before. The future contributions and fresh, imaginative ideas of these young scientists might well serve as bread cast upon the waters to augment and strengthen the Navy's own ocean research and development programs.

From early June to August, our young "Scientist in the Sea" students had cheerfully abided my obtrusive presence and personal philosophy. By dint of considerable effort on the part of the U.S. Navy, NOAA, and the Florida University System—omitting needless reference to our Creator—they were taught all we know about man in the undersea environment. They learned their lessons well over twelve weeks of training, to the envy of those of us who labored twenty years or more to gain the skills and wisdom we now so freely dispense.

During the final three weeks, all sixteen students, man and woman alike and together, underwent a series of saturation dives in Hydrolab, located at 15 meters of depth off Grand Bahamas Island. It was a great experience for all hands, though standing a decompression watch every two and a half days was rough on an old codger like me. As I stood watch in the early hours over Hydrolab, I contemplated that passing along the lessons one has learned is what life is really all about, and I was content with that.

12

FISSH
(September 1975)

During my last few months of active duty in the service of the U.S. Navy (September–November 1975), I was on loan to the National Oceanographic and Atmospheric Administration as senior medical officer for the undersea project entitled the First International Saturation Study of Herring and Hydroacoustics, hence the acronym FISSHH. Since in the end the herring failed to perform adequately, I accordingly shortened the title to FISSH.

My assignment as medical officer to this series of international saturation dives off Rockport, Massachusetts, was a continuation of my ongoing effort to retire from the U.S. Navy. I had planned to retire on 1 September 1975 in a more or less honorable fashion, but then learned it would be necessary for me to repay the Navy the sum of $12,000 for the fourteen days of my contracted services remaining beyond that date. I had never thought my services were so valuable, and certainly the Navy had never paid me at that rate. Had I been judged unfit for further professional services, or court-martialed and discharged for heinous crime, I would have owed only the portion of the $12,000 prorated for the fourteen days. Such were the provisions of my signed contract in the event that I separated from the service prior to completion of the one-year contract. Some legal beagle, however, had ruled that retirement did not constitute separation; I became aware of that interpretation only after I had made my retirement plans.

Shortly, perhaps fortuitously, after I sensibly withdrew

my retirement request, I received a letter from Adm. Ed Snyder requesting my services on the NOAA program, at the end of which I could stow my diving gear and retire. Accordingly, I planned a leisurely trip to Rockport via Bermuda, with Marjorie in tow. I was committed to a series of talks to a crowd of some two hundred recreational divers in Bermuda in August, and my wife and I looked forward to this as a brief but pleasant interlude. It did not quite work out that way. When we found that Marjorie had a basal cell cancer of the nose, we struck new plans. We left our Himalayan cat Lhotse with our son in Bat Cave and took ourselves to Charlotte, North Carolina, where a plastic surgeon performed a nose job on Marjorie. Blessed be! It was apparently successful and I left for Bermuda. Marjorie planned to join me later in Rockport with a beautiful nose and a new lease on life. The cat, however, had disappeared—either stolen or making a solitary trek back to Panama City.

During the talks in Bermuda in August, I seized the opportunity to inform the young divers about every grisly hazard they might encounter in their elected sport or livelihood. I offered to hold an afternoon clinical session to discuss intimately their various ailments related to diving. "I have never seen a truly well diver at one of these symposia," I told them. "The well ones, like the sinners, don't go to church."

I precisely described every potentially fatal problem associated with skin snorkeling and scuba diving, including lung squeeze, underwater blackout, breath holding, thermal control, and bends. As my own concluding film had not arrived by mail, Bobby Robbins, of local dive fame, found a Cousteau film so old that it had not yet been seen by this young audience. In the end I received a rising ovation, in that everyone got to his or her feet and departed. I had to agree once again with Dr. Bob Workman's premise that talking to sports divers on diving safety practices is probably counter-productive. They do not understand that the unfriendly sea does not allow for human error.

In September, I left alone for Rockport, where I started this last Chronicle, based on the sex life of the herring (ichthyological pornography)—an appropriate chapter to close out the book. (The opinions I have expressed, or even implied, in this account are solely mine, and by no means reflect the philosophy of NOAA, the U.S. Navy, the Federal Republic of Germany, Poland, or even that of the USSR. All personalities and events were drawn as I saw them at the time—not necessarily in their true light.)

The Rockport project was a remarkable series of compressed-air saturation dives using an underwater laboratory called *Helgoland* and a multinational team of diver–scientists. The undersea lab, complete with two pressurized living spaces and a decompression room (DeKomRaum) with access to the ocean floor, was placed on Jeffrey's Ledge, about 8.5 miles off Rockport, Massachusetts, at a depth of 110 feet. It was "home in the sea" for many aquanauts of diverse speech and ways of life. As originally planned, the operation was likely to be safe, with assurance of two-way diver communications and regular on-site surface support, although neither of these safeguards came to pass.

The FISSH operation was the deepest compressed-air, prolonged saturation exposure of its time. Unfortunately, inappropriate habitat design, equipment failure, and miserable weather combined to mark this as one of the most medically hazardous extended undersea human exposures in history. Dr. Anthony Low and I shared the burden of medical safety throughout the many days and nights of the exercise, and each of us paid a physiological and emotional toll that will not likely be forgotten in our respective life spans.

The project started off well enough despite a few early language problems that were sorted out as the Americans learned a little German and the German divers spoke English with greater ease. We all learned a curious mixture of the two tongues; however, as we were all divers with common experience in undersea habitats, the problems were not insurmountable. After a few days of organization and joint training, we were underway, with *Helgoland* in tow by the U.S. Coast Guard *Spar*. By noon on 17 September, we were on site, and the letdown of the habitat began. Thanks to efficient coordination, *Helgoland* settled on level keel at 110 feet by late afternoon. Two days later we put down our housekeeping team to accomplish the habitat engineering cleanup. Their brief stay of 48 hours on the bottom regrettably required an hour-for-hour decompression. The chief diver, Hans Belau, remained aboard to head up the first four-man team, the other three being American scientists.

We had less than adequate surface support; no boat was longer than 40 feet keel length, and only two boats were operational at any given time. Of greater concern was the effect of the powerful bottom currents on the saturated divers. From experience in Sealab II, I advised our divers to stay upstream, but this was not always possible. The exceedingly cold water was not troublesome for these divers, who were acclimated to cold-water diving. Should an accident occur, however, our surface support divers and saturated aquanauts alike would be as far as 8.5 miles offshore. At best, this meant a probable transit time of 2 hours before a diver could be placed in either of our treatment chambers, which were back at the Ralph Waldo Emerson Inn at Rockport, as was I. True, we had a small Draeger pot into which a victim could be stuffed for transport ashore, where it would be married to our German treatment chamber, but my aversion to the one-man chambers had been unchanged over two decades, and I prayed ours would never be needed. Nor could we ignore the vagaries of the New England weather during autumn. All of these unpleasant facts were quite apparent to the participants in the program, but were accepted with a degree of equanimity, if not stoicism. The Europeans had faced the rigors of the North Sea and had absolute faith in their equipment. In company with my American colleagues, I was obliged to share that faith.

If there was ever a question of my seniority, the rapid sprouting of my white beard soon laid that problem to rest. At times, I found it embarrassing to recall that I had received one or two college degrees before any of these young divers were born. In fact, my medical degree anteceded the birth of Dr. Anthony Low by several years. Lt. Comdr. Larry Bussey, an old Sealab III team leader, and Gunter Luther, chief engineer for the habitat, were certainly my juniors too, yet I chose to believe the scientific world had not passed me by. Babies are still born in the old way, and medical practitioners make more or less the same mistakes as in my time, or even before my time. Also sporting a beard was Dr. Don Beaumariage, who had not trimmed his growth since we last met in 1972. There is not much to be said for a beard: it tends to itch, and it catches an embarrassing quantity of dribbled food. In brief, I do not recommend growing the beard for young men who have better things to do with their time.

I found our headquarters at Ralph Waldo Emerson Inn quaint, colorful, dry as a tinderbox, and cozy. Just how cozy it was can best be appreciated by a survey of my living quarters, which measured fully 7 feet 5 inches by 10 feet 3 inches with a 7-foot ceiling. I had a bath adjacent with the same overhead clearance. It was not remotely possible to foresee dual occupancy with my wife, even as a momentary encounter, which may well have been the puritanical plan. Marjorie decided to wait a bit to join me, partly to attend our daughter Judy's departure for a year's stay in Amsterdam, and partly to let me settle into the job. We would get together, but not in that room. Our love endured, but our bones tended toward brittle.

Yet, having spent so many years of my life in the cloistered niches of submarines and later in the inner locks of recompression chambers, I found the cubbyhole existence a refreshing atavism. Of course, six weeks is not long in terms of a total life span, especially as I found the food superb and the company great. The gang reminded me of the young, energetic, and immensely talented diver–scientists I had known over the past decades in the Navy's Man-in-the-Sea programs. We spoke a common language, understood the gear we worked with, and shared mutual respect for our individual capabilities. Our West German colleagues numbered about seventeen divers and perhaps a dozen support scientists and surface engineers. Our Polish constituents were almost all Ph.D. level and had worked at least two years with the Germans. The late arrivals, Soviets and Norwegians, were topside scientists, and my American colleagues were all seasoned undersea scientists.

My German medical partner, Tony Low, was a rare jewel in that setting. A soft-spoken young man, he was trained in both aerospace and diving medicine and, after having worked for three years with the diver–scientists, commanded their total respect. When we consolidated our medical supplies and technical libraries, we could easily have established a small hospital in Rockport. In truth, I was happy with the assignment. It was a challenge—the deepest N_2O_2 saturation working dive ever attempted in open sea—but it was also a nice way to spin out an official career.

A few more deficiencies showed up as the days passed. We did not have closed-circuit TV reception from the habitat, and I, who sat a good distance ashore, found this particularly disturbing because visual analysis to me was a critical feature of aquanaut safety. We brought in a TV expert from Woods Hole to correct the situation, but the excuse for the deficiency left me quite cold: Klaus Oldag said that the power intensity for transmission had been designed for 11 kilometers, not 15 kilometers. As I believed that overengineering was always preferable to marginal equipment performance where human lives are involved, I resolved to correct this problem.

Another major misconception of mine from the outset was thinking that we had current meters appropriately emplaced around and atop the habitat, to inform the aquanauts and those of us who monitor ashore of the sea-floor hazards to the diver on a mission. That was not the case. Our German aquanauts were in fact estimating current strength by sifting sand into the local water stream, then using an imported variety of the old Kentucky Correction, made famous by Daniel Boone. Much as I tend to respect American folklore, we had since Boone's time developed Savonius rotors and even better current meters, and they were damn well going to be used throughout the experiment.

At daybreak on the first day of autumn, I assumed the watch, idled up to our control room, and casually asked for the usual parameters of oxygen, carbon dioxide, temperature, and humidity inside the habitat. The response was frightening. The oxygen sensors had all failed shortly after midnight, Hans Belau informed me, and we had no sure way of telling where we stood. Simple arithmetic told me the aquanauts were safe for another 48 hours, unless they inadvertently released some of the pure nitrogen available to them. Still, it was hellish not to have a regular reading of oxygen values within the *Helgoland*.

After a hurried conference with Larry, Gunter, Klaus, and Tony, I began a seemingly endless round of telephone calls, starting with Max Lippitt, shifting to Ed Sharp, followed by two wrong numbers in eastern Pennsylvania, back to Ed Sharp, and finally the welcome voice of Fred Parker, who assured me that new and tested oxygen sensors, packed by himself personally, would be put on a plane and shipped immediately. Not until then did the New England fog lift and some of us begin to breathe once more.

We could have muddled through without the oxygen sensors, but it would have meant starting decompression at once in the habitat chamber, with no next-door attendant in the laboratory, and leaving household engineering chores undone. I would also have bled in large quantities of pure oxygen during the final 40 feet of travel to the surface. Assuming that we got out of that mess with whole scalps, the habitat would still not have been ready for occupation by the first scientific team, and four more days would have been lost.

It could have been worse, however. Yesterday afternoon, 20 September, we had debated bleeding in enough nitrogen to bring the oxygen level down. This procedure would have shortened the engineers' stay in the habitat, and the scientists could have entered sooner; but we reckoned that the engineers were still a little narcotized, so we postponed the addition, for which I thank the Almighty. One can always cookbook a way out of such a mess, substituting pencil and paper and knowledge of physiology for the nonfunctional instruments, but it wouldn't have been a picnic. The volume of the decompression room was much less than that of the other compartments inside the habitat, and calculations could have been quite hairy, with three human lives at stake.

Next I learned that the pO_2 (oxygen partial pressure) sensors installed in *Helgoland* had been designed for monitoring the incubators of newborns, and had not been adapted for hyperbaric use. Under stable pressure—even if elevated—the pO_2 sensors could be calibrated accurately. But when faced with a tidal depth fluctuation of 3 meters, they would lose linearity and become erratic. So late we get the smarts; even then we still tended to calibrate in steady rather than dynamic states. For all hands to see, I pasted on our bulletin board, in three languages, Murphy's Law of the Sea: Any object placed in the ocean by humans will become seasick, lost, or fouled, or any of these three conditions in combination. For most of us, it takes a lifetime to learn that great truth, which also has a corollary: Given sufficient time, any two floating objects, unless restrained, will collide. In honesty, a final rule is in order: It never hurts to know beforehand the people who can help when you are in trouble.

The first of what would be a myriad of crises was thus resolved. There would be many more before Thanksgiving, not all so amenable to slide rule and telephone, and that is why Tony Low and Pappy Bond were on hand.

Throughout the day and night of 23 September, the satisfactory offshore decompression of our two German engineers and Larry Bussey continued. Not so in the adjacent undersea living quarters, however, where a faulty connector resulted in malfunction of the pO_2 sensor, leaving us blind with respect to the atmosphere composition. It was a case of back to the cookbooks. Not a really difficult problem, but complicated by the fact that a fitting blew in the wet room, bleeding in an unknown quantity of compressed air. All of this, however, kept us on the high side, oxygen-wise, in an elevated, but still safe, range of oxygen partial pressure. That we could live with—but not the other side of the coin. The sensor was returned to the surface for repair and recalibration. At the same time, the closed-circuit TV link for monitoring the aquanauts was inadvertently and abruptly terminated. Until these two major mishaps were corrected, we would send no more scientists down to the habitat.

But a great bit of news arrived just before midnight. Lacking much else to do, our aquanauts were taking routine readings of the ambient water temperatures, which had hovered around 6.8° C for days on end. But the temperture had started to climb in the afternoon at the rate of nearly 1° C every 2 hours, reading 10.2° C by midnight. Such a temperature change means little to land-based quadrupeds, but to the herring it is equivalent to a shot of Chanel No. 5, a new coiffure, a see-through blouse, and a set of hot-pants. A sudden elevation of water temperature to the vicinity of 11.0° C puts these little fish into the very motion our experts had been sent to observe and record. Dr. Dick Cooper was with me in Central Control when the report came through, and I would swear that he underwent an emotional alteration closely resembling that of the hitherto unresponsive male herrings. The man became absolutely vibrant, wild to get down there on Jeffrey's Ledge to do his marine biological bit of voyeurism. For me, 11.0° C is still too damn cold for that kind of nonsense.

By 24 September, we were deep into decompression of our first aquanauts, who were resting at 65 feet for a 3-hour period. Tony and I stood the watch heel and toe for this decompression since it was the first for us at this depth and under these conditions. The day broke with continuing rain, some long period swells, and northeasterly winds of about 8 knots. There was no appreciable height of waves, for which we were constantly grateful, as we were well into the hurricane season.

I had spent a very concerned day and night in total ignorance of what damage, if any, had been caused by a hurricane that had swept over Florida. Each time I called the Panama City area, a voice recording informed me that all circuits were out due to storm damage, and we received indirect information that NCSL had been evacuated. After 37 hours of telephone silence, I finally got through to Doris Odum, the perfect secretary and my general factotum in Panama City, who assured me that our populace had survived, although considerable property damage had occurred, especially on the beach. Thorough preparation on the part of various disaster groups had really paid off. At least that one was behind us, and perhaps we'd have a breathing spell for a few years, if hurricane history proved a reliable guide.

At breakfast on 24 September, when some of us put our

heads together to review the situation, a new factor reared
its ugly snout. Cliff Newell revealed that he had been unable
to open the lower access hatch of the two-man personnel
transfer capsule, situated close to the three-man PTC, both
of which provide means of instant escape from *Helgoland* in
event of a catastrophe. Apparently, the 300-pound hatch was
too heavy for one, and possibly even more than two aquanauts could handle. That, of course, posed a hell of a
problem.

Assuming that we could correct the failure of the pO_2
sensor and restore the TV transmission by noon, we had
planned to put two more men in the habitat as soon as the
decompressed aquanauts were picked up on the surface. For
this we needed the small PTC, since we could not stuff four
escapees into the three-man chamber. Accordingly, we set
up a course of action for the day, in order of priorities. First,
the small PTC would be opened, so that all five of our underwater laboratory inhabitants, saturated or otherwise,
could be evacuated if required. Second, the newly repaired
pO_2 sensor would be installed in the main room, and calibrated. Finally, the TV circuit would be restored, a matter of
great importance. The whole process would take time, and
final transfer of personnel to and from the habitat would
have to be accomplished after dark.

Unfortunately, the plan was weather-dependent, and
Mother Nature, so kind over the last fortnight, began to set
our plans awry. The soft breeze of predawn hours grew
slowly to a full-blown nor'easter with winds to 30 knots and
7-foot seas. The only support boat able to handle such sea
states had been designed to take divers aboard in calm seas,
during daylight. To attempt the operation after dark—with
surface currents in excess of 1 knot—would be begging disaster. As a last gasp, Dick Cooper asked if I could massage
Chris Lambertsen's decompression table to get our men out
before dark. I gave it a best try, with all sorts of Bond correc-

tion factors, but could not see a way to surface them until about 1830. In our circumstances, even as late as midnight would have been no worse. Twelve or fourteen hours upstream, perhaps, I could have performed emergency amputations here and there on the decompression tables; but such manipulations simply cannot be done at the tail end of the table. Those who play catch-up in such a situation will lose the ballgame by a very unhealthy score.

The decision was therefore obvious: we would deliver the pO_2 sensor, make certain that the little PTC was opened permanently, and do what we could about the TV. In addition, nitrogen bottles would be lashed to the habitat for the purpose of ultimate dilution of the oxygen levels, long since too high for continuous human exposure. The aquanauts had experienced a dose concentration of oxygen over a period of six or more days at a level that had never been tested, so I put in a call to Dr. Chris Lambertsen for consultation. We were in general agreement on parameters for emergency shortening of our tables, so I set our conclusions to paper and sent it to him for comment. Word from Central Control confirmed that the pO_2 sensor had been delivered and the N_2 bottles as well. Work was underway to open the little PTC, but open-sea approach to the energy buoy was so hazardous that TV reconstitution had to wait. The small Zodiac boat broke its moor and was caught only after a wild chase. If anything, the weather was deteriorating.

Perhaps thanks to my prayers (or else due to a bet of a case of beer with Gunter Luther), the wind had abated by midnight on the night of 24 September and the seas were flattening. With any luck at all, come dawn we would charge ahead. In keeping with the weather changes, bottom water temperatures soon dropped to 7.6° C, which depressed the sex drive of our herrings. That turn of events was dismal to those of us in the field, and probably didn't please the little fish, either.

And so it was that one day ended and another began. My time perspective seemed to have been processed by a blender, coming out as a homogeneous stream of muted crises. Not that I ever complained about the job, mind you. There was, of course, no per diem, and no liberty until Thanksgiving; however, the Ralph Waldo Emerson Inn was certainly true to what we used to say of duty on some submarines: "This boat's not much for liberty, but she's one hell of a feeder!"

Somehow, I had visualized myself in the job as dashing out to sea daily, diving like crazy, and walking at least 20 kilometers every day. Not so at all. I would puff up one and a half flights of stairs to Central Control, stand my watch over the divers, down to the groaning board for a 4,000-calorie meal, on to my room for a nap, then back to Control. That was the cycle, apart from my personal habits and time to read and sleep.

I had learned in a conversation with my wife that she would wait until after her next doctor's appointment before joining me at the Emerson Inn. Her decision was probably wise, as I could not conceive how Marjorie and I could possibly coexist in my cubicle, short of installation of a life-support system. Quite likely, there were larger rooms in the Emerson Inn, but I had not seen them. My room would have been a challenge to any married couple, as to change my socks I had to back into the passageway. God only knew what might come to pass when two very old and great friends met in these confines!

On the evening of 25 September I recorded words I had hoped never again to write. Just before noon that day, Joachim Wendler, age 36, had a fatal diving accident. The facts in the case were clear-cut, and I recorded them with painful clarity.

Joachim, an experienced diving engineer, was one of the three we had put down some days ago to set the *Helgoland* in perfect working order for the oncoming teams of scientist–aquanauts. Another member of the team, Hans Belau, our senior diver, was scheduled to stay on the bottom to work with three scientists who would come down as soon as the housekeeping cleanup was complete. At that point, Joachim and the third member of the engineering team would be joined by Larry Bussey and would enter the decompression room for the 50-hour return to sea-level pressure. They would then ascend to the surface to be recovered by the support ship and returned to Emerson Inn. That was the plan, and it was followed with precision. Larry joined the group in the living quarters after a long underwater survey of the bottom-placed scientific equipment. Shortly thereafter, Larry, Joachim, and the other engineer–aquanaut entered the habitat's decompression room, shut the hatch, and commenced programmed decompression, remotely supervised by Dr. Low and me.

All went quite well, with a smooth decompression and exceptionally good atmosphere control. Tony and I went over each step of the procedure and worked together in the best fashion that I could recall in years. When the decompression was completed, our three-diver team had been stable at sea-level equivalent for 30 minutes. According to the schedule, they were then ready to return to the surface, to be picked out of the sea by support divers on our 40-foot boat *Galatea*. But because the seas were troubled, ranging to 10 feet, and the night was dark, we elected to keep the team in the chamber below at sea-level pressure until daylight and (we hoped) better sea states. That was, I think, a prudent decision. On the bottom, Larry diplomatically cooled the channel fever of his colleagues, and we all attempted some rest.

By daybreak on 25 September—0400 to be exact—the wind had abated, though the seas were still high. Shortly after daylight, the damned wind started up again, and the seas grew hour by hour. The *Galatea* was judged to be the only boat seaworthy enough to go to the habitat site and accomplish a threefold task: (1) put a man on the energy buoy for TV repair; (2) send down ten bottles of nitrogen to dilute the main room and wet room atmospheres; and (3) send down two scientific divers to the habitat, completing science team 1, and retrieve the three divers from the decompression room. The order of these tasks was altered (in itself of no importance) because crew members, arriving on station, could readily see that putting a man on the energy buoy would be too hazardous, so bad was the sea state. They saw further that all hands aboard would be needed to pluck the emerging aquanauts from the sea. In consequence, the decision was made to recover the aquanauts first, then to send Bill and Kenny to the bottom with the nitrogen bottles, to stay below for the duration of their mission.

Accordingly, word was passed to the three aquanauts in the DeKomRaum to make all preparations to return to the surface. In Central Control, we followed each of these developments with interest and approval. The decisions made at sea were correct in light of the very bad sea state and were consistent with the best safety procedures for our divers, always the final criterion. On signal, the three aquanauts in the DeKomRaum quickly recompressed to ambient sea pressure, opened the hatch, and entered the main room of the underwater laboratory. There they hastily suited up, donned diving gear, and exited from the habitat.

Outside the habitat, the three aquanauts began their ascent, following the line that led to the surface "decompression" buoy. Topside, the *Galatea* was tethered close to the same buoy. As the divers ascended, in the face of a 0.5-knot current, they bunched together, giving frequent signals of OK and observing one another closely. Joachim Wendler was carrying equipment in his free hand, but the other two divers were empty-handed.

All three divers reached the surface simultaneously. The

first two swam quickly to the Zodiac, trailing astern of *Galatea*, and were immediately hauled aboard. Joachim Wendler, however, his face mask gone and his mouthpiece missing, raised his hand from the water in the familiar trouble gesture. He gave a loud cry, which could not be interpreted, but there was no questioning on board *Galatea*. Without hesitation, two suited divers went overboard, swam the few yards to Joachim, partially inflated his suit, and towed him back to the boat.

As first seen on the fantail of the *Galatea*, Joachim was not breathing, and no carotid pulse could be detected. At once his suit was cut open, and mouth-to-mouth resuscitation and external cardiac massage commenced in a thoroughly professional manner. No response was evident, but in a wild sea signs of life are not readily detectable, so the effort continued. A call, which we intercepted at Central Control, went out for Coast Guard helicopter aid, and a sudden sense of catastrophe engulfed us all. We could not raise the *Galatea* on radio, but I surmised all hands were too busy for communications, so we prepared for the worst. Tony gathered up his medical bag, our crew went below to line up our two treatment chambers, and we alerted the police and ambulance service to stand by for action.

Thirty minutes later, the Coast Guard helicopter managed to drop two bottles of oxygen on the bow of the *Galatea*, to be used to supplement the mouth-to-mouth procedure. Tony stood by the chambers while I rushed to the harbor wharf to receive the victim. At that point, my hopes for our diver's survival had about gone, but it was no time to throw in the sponge.

At the pier, we were nearly run down by the newly loaded ambulance as it departed for the decompression chamber at the Emerson Inn. Cliff and I were at once put into a police cruiser in close pursuit of the ambulance. We arrived posthaste at the decompression chamber site, where I first saw the body of the victim. He was dead—no mistake of that—but we had to make the final, albeit fruitless, effort. Tony and his assistant dragged the body into the tiny Draeger chamber, and they all went to the ordered equivalent of 50 meters in depth. I could see that they were continuing mouth-to-mouth breathing and external cardiac massage. Quite a few minutes later, Tony answered my question to the effect that he was almost 100 percent sure that Joachim was dead, but could not be totally certain. There were no heart sounds, the pupils were dilated and fixed, and no other demonstrable signs of life were evident. Examination with an ophthalmoscope gave clear evidence of death. But Joachim had been a close friend and ward of us all, so I kept silent while the two live chamber occupants pursued their lost cause, working in a hopelessly efficient manner.

After 23 minutes at the equivalent of 50 meters, I once more asked Tony for an evaluation. He was nearly exhausted, but answered that whether Joachim were dead or not, he wanted to keep up resuscitative efforts. I could not gainsay that loyalty, but reluctantly told him that I would shortly commence decompression. By then, about 3 hours of intensive effort had been spent on a man who now exhibited all the classic signs of death. Further stay at this depth would serve no useful purpose and would put the others in jeopardy, so I started them up. For the next hour and a half they persisted in their efforts at reanimation, until they collapsed midway through the 10-foot stop. In all conscience, I could not ask them to stop, so I clumsily adjusted the decompression table to allow for their exertions and for the inadequate ventilation of the Draeger chamber. Since my operator could not read the tables in the *U.S. Navy Diving Manual* and spoke no English, the irregular manner of decompression was not questioned. Immediately after the chamber surfaced, I entered, examined the body, and declared Joachim dead. At least I could spare Tony that act of anguish.

The next few scenes I would rather not have witnessed. Tony went straightaway to Elske Wendler and told her in quiet tones that her husband was dead. Elske made a futile little gesture of the hands, then covered her lovely face. This reaction lasted only seconds, however; I am sure she had known the truth from early on and had called up those feminine reserves of courage that no man can understand or properly appreciate. After a bit, she asked to enter the chamber, to which Tony consented. A few minutes later, she went back to the Emerson Inn, to call her husband's parents in Germany. The ambulance arrived to take Joachim's body to the hospital morgue.

To put it all too briefly, an extremely competent forensic pathologist performed the autopsy, attended by Tony and me and Dr. John Gale, county medical examiner. The results gave harsh evidence of the initial diagnosis of gas embolism, as I reported to Larry and our group just before midnight. Of course, each one of us asked aloud how this could have happened to a veteran diver. I could only offer a reasonable working hypothesis, as follows.

Joachim was ascending a line tending to an overhead surface buoy, which in turn floated in seas ranging up to 12 feet. He was encumbered with items of gear that he had elected to bring topside, presumably held in his right hand. From time to time, he must have grasped the line he was following to the surface. At some terrible point, he must have breathed deeply at the precise moment that the topside buoy rose abruptly with a sharp wave crest. Hanging tight to the line, Joachim would have been raised about 12 feet toward the surface while in full inspiration. By gross oversight, God gave us no pain sensors in the lung, and Joachim never knew when his lungs reached the critical point of overpressure.

Joachim Wendler died a young man, leaving a young widow and an infant son. For a long while, life would not be pleasant for them. Each one of us was touched by this tragedy. But Joachim died while working at his best-loved trade, in the sea he loved so well. Was there really anything more to say? Life and the program did go on, but that tragic event would live with us all, forever.

By any account, decompression of the three aquanauts left behind in *Helgoland* after the fatal ascent of Joachim Wendler was one of the roughest decompressions in history. Soon after we had completed the autopsy came word that we were probably square in the track of the hurricane that had caused the Panama City disaster. Most of the night of 25 September we spent discussing our alternatives, not one of them pleasant to contemplate. Should the hurricane pass close by or directly overhead, Gunter Luther felt that the energy buoy might well be carried away. In any event, the umbilical was to be disconnected some hours before the blow, thus isolating our aquanauts for perhaps a very long time. If the buoy remained in place, a hookup could survive autonomously for 14 days; but without communications, recovery by itself would be hazardous, and a two-week period in the habitat with zero communications was unthinkable.

At Larry Bussey's request, I went back to the drawing board to prepare my version of the briefest possible decompression schedule, should time permit decompression in advance of the storm. The briefest I could do was about 36 hours, versus the accepted 50 hours, and such an interval would have meant bringing the men up at night, to be recovered from the surface in a fearsome sea state and returned to Rockport on a surface vessel. We abandoned that line of thought, and determined instead to raise the three aquanauts under pressure in the PTC for ultimate completion of decompression on board *Galatea* in the small chamber to which the PTC could be mated. We decided, meanwhile, to begin decompression inside the habitat, and to watch the track of the hurricane. Should it veer to the east, the aquanauts could simply repressurize and resume their tasks. If the storm stayed on track, however, we would use the PTC for evacuation the next day.

As we had feared, the damned hurricane bored straight at us at a 20-mile-per-hour clip, with no inclination toward England or even Greenland. The decision was inevitable. We enlisted once again the assistance of the Coast Guard, this time to help us retrieve a bouncing capsule loaded with half-decompressed aquanauts, then to take them ashore to our Rockport chamber.

Tony by that time was totally wiped out, having survived the long ordeal in the chamber with Joachim the afternoon before, the postmortem exercise until midnight, and the skull-session at Central Control. It was logical to leave him ashore for a few hours of sleep, while I went to sea to oversee the lift and recovery of the PTC with its occupants. By the time the PTC had been delivered to Rockport and a satisfactory mate made with the Draeger chamber, Tony would be rested sufficiently to take the first of what would be many decompression watches in the chilly circus tent that encompassed our two chambers outside the Emerson Inn.

We finally got underway aboard the *Galatea* in moderate seas. About an hour later, we made rendezvous with the Coast Guard vessel *White Heath*, circling on station in the vicinity of the habitat and its many buoy arrays.

Maneuvering the vessels to recover the wildly bobbing capsule with our aquanauts inside was a highly intricate procedure, demanding complex communications crossfire, difficult and hazardous navigation, expert seamanship, and superb deck handling for a matter of hours. We finally got the PTC with its human cargo aboard and secured in position by late afternoon. The hasty return to Gloucester port was anticlimatic, as was the wild ride with the capsule aboard a flatbed truck and the eventual mating maneuver with the Draeger chamber. At that point, I turned the bit over to Tony and retreated to a hot bath and sleep.

The last stages of decompression began at Emerson Inn and stretched through another 40-odd hours. For the inhabitants of the Draeger chamber, the interval was eons of misery, but they were professionals and all the gripes I heard were in the best of good humor. Despite his attitude, however, we were concerned about Hans Belau, who had breathed an atmosphere containing more than 0.7 ATA (atmospheres absolute) of oxygen for more than 200 hours, longer than the others. Tony and I watched him, awake and asleep, like hawks, and judged it safe to complete his decompression in a routine manner. Following a somewhat shortened decompression table that I had improvised, the aquanauts were surfaced. After two hours of intensive physical examinations, Tony and I took them to the Gloucester hospital for repeat hematological examinations. All of them were well, and we were ready to go again—admittedly, it was not the easiest of lives.

By the next day our surface support divers were busily planting their fish traps and repairing the damage to the energy and decompression buoys. The weather was perfect and the sea calm. Slowly but surely, we prepared for the second team's occupancy of the undersea laboratory. On the day after the PTC was emplaced, the team would go down.

We had a farewell party for Elske Wendler. I think it was for Joachim Wendler as well. Larry Bussey played the grand piano with gusto and finesse, and we all sang a lot of very old songs. Of course, I needed no lyric sheets, since memory served me well; but the young folks who sang along had to follow the written words. Two of our German friends joined in, one with a four-stringed banjo, the other with a guitar. My harmonica accompaniment lent little to the concert, but I gave it my best. Elske seemed to appreciate the show; at least she did not break down. She was a very brave woman. Although we felt deeply her sense of loss, she would survive the catastrophe. Tomorrow would always be another day, bringing its own surprises, happy or otherwise.

13

FISSH
(October 1975)

The first of October was a truly beautiful day. The sea was calm, the air was fresh, and the maple and ash trees had begun their annual display. Our divers put finishing touches on the habitat, surface buoys, and associated array. The PTC was repositioned, and we were ready to put the next team of aquanauts on the bottom as soon as a final OK was received from Washington. I took team 2 and the remnants of team 1 to the hospital for the required blood examinations. By now, the hospital personnel knew me by name and countenance, yet I sensed a certain air of professional distrust. (Their attitude might have been related to my scraggly beard and unconventional attire, but I chose to believe that people in the Commonwealth of Massachusetts could not conceive of a bona fide practitioner of medicine below the Mason–Dixon line. Of course I had been setting bones, removing pieces of gut, and delivering babies long before these young Yankees were conceived.)

We had received official word that the international saturation study could go forward after some necessary tightening of loose seams. Admittedly, some reorganization was in order. The German group had its own internal organization. One of the group was an expert in all matters pertaining to the habitat; another was qualified in the communications network. A third best understood the various decompression chambers, and finally Hans Belau, senior diver, was in charge of all diving exercises. In theory, all of these directors reported to Gunter Luther, who was himself not a diver, as was the case with his executive, Klaus Oldag. Under the procedure followed in Germany, each director handled his piece of the action independently, with a rather loose program of coordination.

In U.S. territorial waters, however, the system had needed tightening. Under the U.S. Navy system of organization, the on-scene commander, Larry Bussey, who reported to his NOAA senior, was responsible for safe and successful conduct of the overall operation. Our German colleagues viewed this as merely a pro forma arrangement at best, which led to occasional breakdowns in our lines of communication. Problems that pertained strictly to the habitat, the energy buoy, or the PTC often did not filter up to the highest level in a timely fashion.

A good example of such lapses had occurred when we raised the PTC and its occupant aquanauts, under pressure, in the onslaught of the late-September hurricane. Just before communications with Central Control were terminated, Hans Belau had requested permission to return the internal pressure of the PTC to 32 meters (ambient bottom depth) after initial overpressure to seat the hatches. The message was relayed to me on board the recovery vessel, and I quickly gave approval. Ten minutes after overpressurization, I noted a mass of bubbles on the surface, and concluded that Hans was simply venting to the agreed-upon 32 meters. Fifteen minutes later, when the PTC was midway up on her journey to the surface, I was surprised to note a second long burst of bubbles. When at last the PTC broke surface, a third long vent of gas was audible. I sent a diver over immediately to read the top-mounted gauge, and he reported a figure of internal pressure that factored out at a disturbing 26.5 meters. The level was safe enough at that stage in the proceedings, but we had no voice communication with Hans, and I had reason to believe that he had resumed decompression on his own tables, and was probably planning to return to the depth of their last stop before leaving the decompression room in *Helgoland*.

Since the aquanauts had spent 1 hour and 40 minutes at bottom pressure, and another interval after leaving the 21.0 meter depth of decompression, this rapid decrease in pressure would not do at all. Accordingly, I ordered Manfried Pose to secure all external valving on the PTC, to prevent further decrease in pressure. Some 30 minutes later, when communications were established, Hans wanted to know, What the hell? I told him rather abruptly that they would stay at present depth until they were safely inside the Draeger chamber ashore, when I would resume the decompression. This decision was accepted, and it was so done.

Two days later, I had sat down with all parties concerned and explained why I could not, for decompression purposes, simply disregard the 1-hour, 40-minute return to bottom pressure and more; nor could I count it as an excursion dive for calculation, since they were not equilibrated at 21.5 meters, but rather were supersaturated. As gently as possible, I further told them that once they left the *Helgoland* decompression room, the control shifted to Dr. Low and me and to no one else. We came out of that meeting with mutual understanding, and I was sure that problem would not arise again—at least, not on my watch.

That singular incident might be rated by some as a minor event. But unplanned things that happen when you have saturated divers under pressure are never minor in my book. Laughable, at a later date, perhaps; excusable, under some circumstances, or of no immediate concern to the pressurized inhabitants; but never did I perceive them as minor. We had lost forever one highly trained aquanaut through what may have been a minor error of judgment on his part —a very sobering reminder to us all that diving is not a safe profession.

By the beginning of October, the reorganization had

given me the new responsibility of senior diving safety officer (on site). Morgan Wells was my headquarters boss, and Tony Low my deputy. The assignments suited me fine. I had known my boss for many years, and had secretly yearned for a deputy ever since I had made lieutenant. Lines of communication and responsibility were clearly defined.

After reading everything I had on hand, I had discovered some remarkable titles in the voluminous library at Emerson Inn and had read down to the last word *The Last of the Great Scouts*, by Helen Cody Wetmore and Zane Grey. I enjoyed the book, although I must admit that my interest was somewhat parochial, as I had attended Riverside Military Academy in North Carolina forty years before with Buffalo Bill's grandsons, and Zane Grey had been a fishing partner of my father; I remembered the author's visit with us at Long Key.

Late evening on 5 October, good news finally arrived from the support vessel *Flying FISSH*. Full electrical power had been restored to the habitat, a development for which we had been waiting before sending down team 2, and her high-pressure air flasks had been recharged. TV service would take another week, which I found disturbing, though I decided to live with the delay. In any event, the herrings didn't care, since they had cautiously commenced to spawn, much to the delight of Dick Cooper, who went down to be there in their midst.

Atmosphere control was finally achieved in the underwater laboratory, with the pO_2 holding at 0.32 ATA, which is where we wanted it. All other factors were good except the current, which measured 2 full knots. But our men were professionals, trained to handle the situation, and it certainly did not seem to worry the herring.

At that point, however, I became quite concerned about the exposure of team 1 to oxygen levels that had ranged from 0.68 ATA to 0.82 ATA for days on end. Of special concern was Hans Belau, who after tolerating this insult for more than eight days, had then endured more than 60 hours of decompression at high pO_2, with terminal exposures of 2.2 ATA of O_2 for periods of up to 1 hour. He survived the experience with a clear set of lungs. Even more remarkable, neither Hans nor his team partners experienced a statistically significant drop in hemoglobin, hematocrit, or any of the common indices of red cell evaluation. One of the team showed the expected dip in thrombocytes, but nothing dangerous, and would have his third and final hematologic evaluation the next day. Of course, we hadn't completed statistical analysis of our scant data, but all divers of the other teams would be followed closely. As we had expected, all team 1 members lost weight, ranging from 3 to 5 pounds.

Although some narcosis must be expected at the 32-meter mark, it was not evident in our frequent and prolonged technical conversations with the aquanauts. Bottom water temperature now hovered around 10° C, and we had no complaints from below. Habitat temperature was held at about 23° C, with a relative humidity of 74 percent. CO and CO_2 levels were extremely low.

Near the witching hour I made a final check with the underwater laboratory. The aquanauts below were my wards, as would be the teams to follow, but I relished this watch as my last undersea adventure. I had other wards, neglected over the years, and it was long past due that I recognized that fact. The natural span of life is finite, and it was time for younger and smarter diving doctors to stand a few late and early watches.

On 6 October the weather prospects were not promising. Rising winds were predicted to reach 20 knots, with 8-foot seas near Jeffrey's Ledge, and much work remained to be done. As was so often the case when the air–sea interface threatened the safe employment of surface support divers, we turned to the bottom-dwellers for assistance. If current and tide permitted, the aquanauts would assume the work-load without a murmur. Nonetheless, their scientific program suffered for every minute they spent mule-hauling on the bottom.

A curious dichotomy widened daily between aquanaut divers and those topside who had never been at the mercy of the sea. For perhaps the fifth time, the down-haul manually operated winch system for the material-delivery pot failed, thanks to poor design and assembly. When the aquanauts complained mildly, they were told to throw away the rachet and wheel and haul the damned thing down by hand. Within the next week, a battery was to be supplied to power an electrically driven winch. Meanwhile, more time and effort extracted from our aquanaut–scientists, who have more important things to do. Perhaps I'm biased, but I believe that topside bureaucrats should be qualified divers, just as firmly as I believe that proofreaders should complete at least three grades of elementary education.

The ancient architecture of the Ralph Waldo Emerson Inn, which lies somewhere between neo-Gothic and late Italian Renaissance, was cleverly designed to sound like a steam calliope in even the mildest of summer breezes. On 9 October we had a Beaufort Scale 6 coming off the ocean, and banshee wails emanated from every corner of my little cubicle. Up in the Central Control for a predawn briefing by our aquanauts, I learned that a couple of the divers were 150 meters on a transect line southeast of the habitat, covertly watching the romantic convolutions of herring couples. Shortly, they would return to the habitat, chock-full of erotic ichthyological minutiae to share with other marine biologists.

Having reservations concerning the accuracy of our oxygen sensors, which seemed to be reading on the high side, Tony Low asked Dick Cooper to do a rough calibration on one instrument, which he did by flowing a small stream of compressed air across the face of it. The reading came out 0.72 ATA, too low a level in view of the peak high tide at the time; the reading, at 36 meters, should have been around 0.86 ATA of oxygen. I was distressed by these results, and ordered that one of the sensors be sent up to me on the next day. Using a more refined technique of exposure, I planned to make tests with 100 percent oxygen at sea level, then take the sensor down in the chamber to 150 feet or so, recording during descent. Based on the readings, Tony and I would construct an accurate calibration table for both our own use and that of the habitat occupants. I prayed that the aberrations of the instrument would prove to be linear; otherwise, we would have big problems and a lot of kitchen-table physics ahead of us.

I recalled the dim glory of Sealab days, when we piped our gas samples continuously to the topside monitors, where accuracy was never in question, and certainly not subject to instrument drift because of tidal pressure effects. Later experience had shown that we could ride on the high side of oxygen, and I was prepared to go the cookbook route. In the case of FISSH, however, this approach would be merely a measure of expediency to complete the mission of team 2. Six more weeks of such a mathematical game of chance would be unthinkable.

We were advised early in October that U.S. Senators Lowell Weicker and Milton Young, complete with a full retinue, were scheduled to arrive on our scene soon. Since each of these elected representatives had been photographed wearing scuba gear, they planned a visit to the *Helgoland* habitat. I wished to God they would just ask for a written factual report of our program, rather than insist on an onsite survey. We appreciated the honest interest of high-level legislators; we nurtured their embrace of all projects such as ours and stood prepared to cooperate in an extraordinary fashion, yet I had qualms about their proposed in-depth in-

vestigative visit. True, when Senator Weicker had partici-
pated in a saturation dive with the Hydrolab habitat in the
Bahamas waters some months ago, he had acquitted himself
very well. But our project operated in cold, rough seas, at
much greater depths, outfitted with a totally different set of
diving gear. The combination took getting used to, and we
wouldn't have much time. I could only pray for perfect
weather, a satisfactory checkout prior to the dive, and great
skill on the part of our selected buddy divers. As bottom
time would be limited to 12 minutes, visitation privileges
would be confined to making a circle of the habitat. Time
permitting, the senators might stick their heads up in the
access trunk for photographic purposes, but no more.

I had not yet informed the aquanauts of the impending
visit, since they had already undergone more than their
share of fits, starts, and alarms. Above all, they were scientists, working day and night to do a difficult and dangerous
job. Public relations were important, and congressional
funding even more so, but for this visit I would not disturb
their work routine, nor cause them undue anxiety.

In a conversation with Marjorie, I learned that another
birthday of Valley Clinic and Hospital, founded in 1948, had
been well celebrated and that my mother, age unknown, had
been present in a wheelchair. The annual event meant a
great deal to her, as indeed it did to all of us who wisely
invested our efforts for the hospital's existence. A lot of lip
service has been paid to rural health delivery, but our hospital stood as living proof that it could be done—without
federal patronage.

As a happy afterthought, Marjorie also told me that she
would visit me in Massachusetts in late October, after a last
inspection by her surgeon. At the Emerson Inn's going rate
of exchange, thirty-five clams per diem, we will barely have
time for a front door embrace, and I have advised her to
pack a very small valise. On the ragged edge of retirement,

one must be quite careful about these matters. The last bit of good news was that Marjorie had sighted Lhotse deep in the midst of a "Laurel Hell" above our house in Bat Cave. I reckoned that, within a month or so, the wayward feline would either return to the human fold, else be forever lost in the mountains of North Carolina. In any event, I wished the creature well. We all loved him dearly, but he had his own secret life to live, as do we all.

Come December, by decree of the laws of the land, I was to retire from active duty. That would surely happen, I was told, whether the herring spawned or not; indeed, if these fish were to cop out in their procreative drive, there would be small reason for my presence, save to amuse the natives with my southern manner of speech and unspeakable beard.

In any event, I planned to make a final appearance at NCSL on 1 December, somewhat overweight in my blues, but wearing the requisite medals and saying the threadbare phrases. A few days before, many of my old friends and perhaps a handful of my enemies were to gather at the outskirts of Panama City to wish me smooth sailing or bad cess, as the case may be. I was sure to be moved by this, but I believed I would echo the sentiments of Napoleon Ledbetter, who was tarred and feathered and ridden out of Polk County, North Carolina, for voting Democrat. His parting remarks are indelibly fixed in my mind: "Except for the honor of it, I'd sooner walk."

14

FISSH
(October–November 1975)

About six centuries past, my favorite poet, Geoffrey Chaucer, penned the following lines (frequently plagiarized):

"The lyfe so shorte / The crafte so hard to lerne."

Chaucer had a valid point, particularly in the craft of undersea activity. Moving an hour at a time through the FISSH project, by 10 October I came to have an inescapable sense of déjà vu. Humans being creatures of habit, I suppose that they are doomed forever to repeat their own mistakes and those of their predecessors as well. On dry land, their mistakes might lead only to fender-benders and smashed thumbnails. In the undersea habitat, it was a more serious matter.

Earlier on 10 October, in a long chat, aquanaut Dick Cooper had given expression to the same thought, stated with more elegance by sundry politicians, historians, and philosophers: Those who refuse to read the history of the past are forever condemned to repeat its mistakes. Apropos of this, Dick and I made a simultaneous observation that the volume of effective scientific work performed by the *Helgoland* aquanauts was inversely proportional to the number of visits by topside diver–assistants and to the volume of radio traffic from Central Control. (From dim reaches of memory, I recalled that in 1962 one of my junior medical officers did a statistical study on our chamber subjects in Project Genesis, plotting word counts, inside and from without the chamber, versus the volume of useful work accomplished by the chamber occupants. There, too, inverse proportionality was demonstrated. Perhaps not unexpectedly, his two curves crossed on about the fourth day of incarceration, and diverged expotentially from there.)

This basic philosophy of inverse proportions I gently expounded at our meeting that evening for all FISSH hands, and agreement was universal. Voice communications with *Helgoland* were to be lessened and surface support dives minimized. Quoting Dick Cooper, I raised the question of who was supporting whom in this operation. We all respected his philosophy: The aquanauts were not to be disturbed by endless communications, nor upset by needless alarms. They had a professional mission to accomplish, and our job was to assist, not hinder, their work.

Later that night, in the course of technical discussions, Dick Cooper, "Atka" Pose, and I reached agreement that oxygen levels in the habitat (0.43 ATA), although higher than desirable, would nevertheless be allowed until I had opportunity, a few hours later, to calibrate one of the habitat's three oxygen sensors in our recompression chamber. The instrument was to be sent up in an early morning pot transfer and the results radioed to the aquanauts. I had considerable, but less than devout, faith in the accuracy of the devices. A calibration run would cost me only about an hour in the chamber and would eliminate, for a while, our cookbook maneuvers.

Marjorie had been expecting to join me in Rockport on or around 16 October for a brief stay. Not only was that the time, unfortunately, that Senators Weicker and Young were due, but it was also our target date to start decompressing team 2. If the decompressions ran the way they had in the past, Marjorie could occupy my little Emerson Inn cell undisturbed (and unvisited) for four days, which is the limit of our dwindling cash resources. The original idea was that we might see one another for brief periods, but the revised schedule held no promise. Accordingly, I asked her to delay a week. Senators and decompression schedules were not to be denied.

When I realized that I did not know or care whether it was terribly late that night or quite early the next morning, I put my watch aside and thought in relative time. How many days and hours into the mission? How soon would the tide be high? When would we get maximum bottom currents on Jeffrey's Ledge? How long since we had changed canisters in *Helgoland*? How long had our divers been out on the 200-meter traverse? Those were the only meaningful measures of time as I assembled the minutes and hours of each span from one to the next. It was as good a way to go through life as any, although it didn't seem very orderly.

On 12 October I took a day off to go to Frank Scalli's, about 2 miles away. The occasion was a surprise party honoring the great *National Geographic* underwater photographer Luis Marden, whom I had known for about twenty years. Many of my old friends from New England would be present, and I looked forward to the affair. Seeing acquaintances after a span of so many years didn't fill me with any particular dread. The faces, voices, and mannerisms I had known so long ago would still be there—I needed only to seek them out under the layers of time.

With all this in mind, I made a final check with Central Control and Tony Low, who would cover for me, then bathed, shaved to the outer perimeter of my ghastly beard, and donned what might pass for civilian clothes. Since my colleagues and the hotel staff had never seen me in mufti, an unmerciful bit of razzing ensued.

There were, of course, snide remarks that the celibate life had finally become too much and I was probably off tomcatting. I tried to quash that slur with a quiet remark that poor Aunt Elizabeth had at last passed on, and I must be among the handful of mourners at a simple graveside ceremony. Behind the transparency of that hastily concocted lie, I was wearing an aloha shirt, green slacks, yellow socks, and white goatskin shoes—not to mention my natty houndstooth jacket. I sniffed as disdainfully as possible and strode out, head held high. The last remark I heard was from my favorite waitress, and it cut deep.

"Hey, girls," cried Gretel, "Pappy even smells good!"

The party was fantastic. Frank's house dead-ended on Linwood Place, so my Department of Commerce carryall vehicle was shortly boxed in by vehicles belonging to other well-wishers. About seventy-five of us celebrants were seated in a large candy-striped circus tent, from which we did not move for the next 3 hours. (Well, there were occasional departures to a portable relief station rented at considerable cost.) We remained a remarkably cohesive group, screaming at old friends across the long tables, and screaming at our table partners at about the same decibel level. I was lucky enough to be seated between the lovely Dr. Eugenie Clark and Dr. Harold Edgerton.

As soon as we had all taken our places, word was passed that Luis Marden was at hand. We dutifully quieted down, then gave him a standing ovation when he stuck his head through the tent flap. Luis was flabbergasted, since he had for months been led to believe that the party was in honor of Harold Edgerton, and that he himself was to be master of ceremonies. In a state of shock when led to his throne (borrowed from a museum for the occasion), Luis without thinking began to recite the glowing phrases he had so carefully memorized to pay tribute to Edgerton. Midway, he recovered with great aplomb, then went on to move us to loud applause and tears.

Then came the feast: clam chowder; pecks of steamed clams straight from the beach, served with hot, puffy clam cakes; lobsters in unbelievable quantities. Personally, I disposed of two good-sized lobsters, and word has it that others were even greedier. The wine was excellent and enhanced the meal perfectly. In all, a wonderful 3-hour orgy of traditional New England clambake and conversation. But as my royal carriage was scheduled to become an evening duty–watch pumpkin, I took hasty leave of my many friends and left to sort out my government vehicle from the hundred or so parked between Gloucester and Rockport. Upon return to home port, I shucked off the clothes of civilian indulgence and checked in with Buzzard's Roost (Central Control).

It had been a quiet day. The hot water heater had failed; there was a minor leak in the decompression room; the pO_2 level was still a bit high; and the habitat tended to rock to and fro with each overhead swell. In all respects, it was a comforting report. At sea, one tends to settle for less than the best of all possible worlds. Our aquanauts were alive and well and in no immediate danger that we couldn't handle.

Outside, the wind had backed to the northeast and the seas were troubled and high. We would not put divers down during the coming day, but a few faulty valves and the thermoregulator for the hot water system could be floated up for repair or replacement. At this point my good and respected friend Hans Belau made a remark that I translate with a free hand: "I cannot understand why, after more than five years of faithful function, we are beginning to see so often failure of our reliable components."

Poor Hans! It was almost as if he expected that if system components had been reliable enough to survive five years of wear and tear, they would, therefore, last forever. We do not find it so with the human body, and I have never seen it so with man-made appliances. We should long since have compiled an elaborate list of system component failures as they occurred in habitats over the past decade, submitted them to elementary analysis, then made up our shopping list accordingly. In the cars we drive, most of us have an instinctive knowledge of what will wear out first, and we plan accordingly.

By 14 October the seas were too high for support boat activity over *Helgoland*, and all hands instead fell to work on the little things that make a project go as best it can. Lines were eyed and spiced; a heater system was jury-rigged for the decompression chamber; hoses and lines were made up for connection to the habitat; safety manuals were revised; and Wolker, perfectionist mechanic that he was, worked all day on his newly acquired secondhand Buick. In this process, under my expert guidance, he thoroughly soiled his orange jumpsuit. I left him, lying beneath a dripping crankshaft, to put my laundry through the washing machine before he could foul the appliance with axle grease and spent lube oil, as he does almost daily. The usual result of our consecutive wash cycles is that we both wear jumpsuits fit only for grease monkeys. In truth, I fear that the splendid washing machines of the Emerson Inn are forever contaminated with the grosser elements of petroleum degradation.

The evening routine I had followed so far into the project generally started with dinner (at Emerson Inn I dare not call it supper), followed directly by our nightly directors' meeting, where we reviewed events of the day, discussed the next day's objectives, and agreed on assignments. After the meeting, I would go to my tiny room to read and nap, then visit Central Control for an hour or so before midnight. That was a good time to talk to the aquanauts and mull over fine points of philosophy or strategy with Larry Bussey, Tony Low, and other key people. By then, it might be well past midnight, and I would return to my cubicle to document the day lest some worthwhile points be forgotten.

The setting of the evening routine changed a bit when, of a sudden, whatever powers may be decided to move me to new and sumptuous quarters in the Emerson Inn. My entire load of personal effects, previously stowed in a space roughly equivalent to that of a VW van, was moved to a room thrice the size. A corner room, no less, with a magnificent view of the cove and harbor. Within minutes after I had unpacked, my new quarters were just as jam-packed as had been the case in my solitary cell. For damn sure, no vacuum existed there, so nature, at least, would not abhor my living habits.

By that time, our aquanauts had made their last scientific sortie, and for the next several days and nights we had to deal with the mechanical, logistical, and medical problems of getting them safely once more to sea level and ashore— the third time that one of our teams would undergo the decompression cycle. Each of the preceding cycles had been radically different in management and outcome. As a matter of simple odds, we should do better this time, but as optimism was not my stock in trade in any conflict of man against the sea, we tried to be prepared for all worst possible situations.

The full senatorial deluge was expected on board by early morning on 18 October, several days away, exactly the time when we would be decompressing Dick Cooper's team and starting the next saturation group on their mission. Our visitors would expect a physical clearance, a shallow 10-meter water checkout in the cove, then a dive to the habitat and a quick bottom survey. We had pleaded for a two-day session with our visitors: the first full day for pool and shallow-water dives; the second for a fast inspection dive to the habitat at sea. Some of the dive candidates had never worn wet suits, all would be diving with somewhat unfamiliar gear, and the time that had been allowed for proper evaluation of their underwater capacities was simply not enough. The situation was made more tolerable when an additional, very competent diving medical officer was assigned to supervise the congressional dive activities. To a considerable extent, the officer would ease the load on Tony and me, but the ultimate burden of responsibility would, quite rightly, lie in our laps.

We were, we protested, the best technical team they could find. Then why were we not allowed to draw the parameters for the congressional undersea exercise? It was unbelievable that a U.S. senator could dictate to me when, where, and how he and his neophyte staff would dive on our site. If one of them should be injured or perhaps killed, I—not he— would be at the wrong end of the long green table. In my book, it was a damned sorry way to run a railroad.

We continued to march along at the accustomed snail's pace in decompression, but by 18 October, the day of the senatorial visit, the wind had started to pick up and we heard a predicted velocity of 35 knots by midday. If it served no other purpose, it might at least deter our congressional delegates. As the day wore on, the seas began to mount, and by noon, as predicted, gale warnings were up. As long as the wind did not veer from the northeast, prospects for any surface support activities were squelched. As a sort of sop to congressional pride, I offered to take Senator Weicker to 132 feet in the decompression chamber, just for the ride. My motive was not altogether altruistic, as we had some depth gauges that needed calibrating, and I hadn't ridden the chamber for almost a week.

I left the watch to Tony and tried to get some sleep before the chamber dive, which was planned for early afternoon. The howling of the gale soon awakened me, and some minutes later we had a massive power blackout in Rockport. We scratched the congressional chamber ride, and Tony and I continued monitoring a blessedly uneventful decompression. The weather, if anything, had worsened.

On 21 October we experienced a nightmare that not many of us were likely to forget; in sheer scope of potential tragedy, I could scarcely recall its equal. Several hours after midnight, I was talking with Dick Cooper on the radio static circuit when he told me of a severe underwater shock, either explosion or implosion, that had rocked the aquanauts in their beds some 10 minutes before. A few minutes later came the frightening report: "Pappy, the lights have gone out. We've lost all power." That started my cage a-rattling.

"The lights are back, but they're going fast! Now they're gone, and we're on emergency battery."

My own lights went out and the radio communication became garbled. We were experiencing an underwater earthquake.

On the bottom we had four aquanauts who were decompressed and ready for return to the surface come daybreak. In the main room of *Helgoland* we had aquanaut Nicholaus alone under pressure, unable to troubleshoot the mechanical problem because no aquanaut was allowed to go outside without a diving buddy. Surface support divers could not be delivered in less than 6 hours' time, and even that help was contingent on the sea state, which showed no promise of improvement. The situation, at that predawn hour, was not good.

We awakened Klaus Oldag and informed him bilingually of the multiple problems, since his precious energy buoy was a key factor in the current catastrophe. He talked at some length about how he would correct the problem of dirty oil and install new oil filters after he was put aboard the energy buoy. It seemed impossible to put the man on that treacherous hulk in such a sea state, so I dismissed the proposition and began to think of the unpleasant alternatives.

Tony, Klaus, and I faced several plans of action, not one of which was palatable. Assuming that the *Helgoland* could function apart from the energy buoy for as long as one month, as I had been told, there should be no problem. Our aquanauts would simply sit tight at their various depths and await more clement weather, when surface support would solve their external mechanical problems by making repetitive dives. But the solution proved to be not quite that easy.

In rapid order, I was informed that an external fitting leak in our oxygen system had dissipated the possibility of further oxygen use during decompression. Since my four aquanauts were so near sea-level equivalent, only another massage of Lambertsen's tables was called for. I was concerned, however, about the separate oxygen supply that was reputedly to be made available for metabolic needs. Shortly, I was disillusioned on that score. That system, too, had developed a leak and had lost all the gas. Although the oxygen requirements to sustain human life could be met by frequent ventilation from our compressed-air banks, the banks would not last long because our energy station was out of commission and we had no other means of gas replenishment.

At daybreak on 4 October came a request from Dick Cooper that the aquanauts repressurize the DeKomRaum to 32 meters in depth, then suit up and go outside to correct the mechanical problems. I agreed with some reluctance, since such a procedure would obviously distort my basic plan of decompression. The logic, however, was clear, so I gave my consent and set about once more to modify our decompression tables.

The aquanauts completed the problem-solving task within 18 minutes from portal to portal, and from that point on I was committed to an escalating game of chance with respect to the ultimate decompression schedule for that group of divers. They must have realized the extent of my uncertainty, since each diver who embarks on a renewed descent into the water or pressure column changes the rules of the game, according to the time interval owed to attain sea-level pressure. When a decompressing saturation diver interrupts decompression to dive again, all bets are off.

The aquanauts had been largely desaturated, but incompletely so. As they had then made a prolonged return to the 32-meter depth, I had to recognize a recurrent saturation of certain body tissues, and thus had to start anew the computation of decompression time, in accord with the Haldanian concept. I preferred to believe, however, that all tissues of the human body seek an equilibrium of gas tensions, and we mortals could only guess at bits and pieces of the mysterious gas transport puzzle. Lacking better wisdom, I settled for that.

But I was enraged that the sacrosanct German machinery had shown such a lousy track record. We had come within minutes of losing or seriously harming our aquanauts, thanks to hardware failures. That night our meeting turned into a real knockdown battle that was inevitable because American and Germanic philosophies were clearly at odd's ends. I became especially enraged when Klaus Oldag reprimanded me for not stopping the aquanauts' evacuation process once he had accessed the energy buoy. Impatiently, I explained that once recompression is started, there is no choice but to continue the process and to complete emergency evacuation as rapidly as possible. Even had I known that Klaus had managed to get aboard the energy buoy—and we were not so informed until recompression was nearly complete—I would have had no confidence that he could restart the diesels and restore air and energy to the habitat. We were at the time dangerously low on stored compressed air, vital to the escape of our aquanauts from the DeKomRaum, and the system was leaking. Larry and I had made the decision to get the divers out, with a secondary plan to evacuate Nicholaus via PTC 1, and that was that. Klaus and I could not agree at all on what had been done; a message to call Panama City at once got me out of the room just in time.

The four aquanauts were at last brought safely to the surface, without any physical damage discernible to my anxious eye. I escorted them to the hospital for blood tests and had pulmonary function tests done as well, all of which came out normal. I ended that decompression period contemplating the biblical quote that I invoke so often and understand so poorly: "Sufficient unto the day is the evil thereof." To me, it has always seemed to mean, "Wipe the board clean and start a fresh record." Or as a more modern wit might put it, "Cheer up, today is the tomorrow you worried about yesterday!"

At the 22 October evening meeting, I still wanted to wring Klaus Oldag's neck, but that primal urge had been sublimated to a third level of consciousness. As we say in medical terms, it was reduced to the hypothalamic level. But it was ready, awaiting only my beck and call. When three of our American aquanauts de-volunteered from the FISSH habitat program, I could not find it in my heart to blame them. They promised to continue service as support divers, but they had lost faith in the safety of our program vis-à-vis saturated divers. Quite likely, a single fortuitous incident or apt sea tale might have been all that was needed to get our professional troops back to battery. Search my mind though I did, I could not find a single word or act that would turn the situation around. The possibility of an aborted mission was very real indeed.

Then, near midnight, came a blaze of new intelligence that opened all our eyes and probably turned the tide. Klaus Oldag recalled that when he had mounted the energy buoy, under most hazardous circumstances, the entry hatch had been wide open. Not only that, but a bird that had somehow entered the air intake tube had been defoliated and died therein. Almost instantly, the problems of the malfunctioning energy buoy were solved, and my apologies went out to Klaus. Beyond doubt, an inquisitive seaman had mounted the energy buoy and entered it, leaving the hatch open when departing. It would not have been long before salt water drenched the TV transmission system. To cap it off, the small bird had flown in and was sucked into the intake system of our support diesel engines. One could only point to an act of God. Although we would have loved to find the fisherman who left the hatch open, the important thing was that we had closed our ranks. I knew that the job would be done.

By this time, I had become acutely aware that Tony was fatigued to his utter limits. He was remarkably sensitive and professionally capable, but his physical breakpoint could be measured in hours, not days. In my own case, it was somewhat different. I had lived a great many days, not a few of which had been laden with grief. I would not wish to say that anyone can become hardened to loss of life or even to the emergencies that arose in the course of operations such as ours. Rather, it becomes possible to scan the horizon for the longer reach of good fortune, and to swallow the tears that always come. The fight was never easy, and it had been especially hard on Tony. In due time, he would toughen himself to these darts and arrows. Meanwhile, I was glad to be on hand. The mere presence of a grizzled veteran does help, sometimes.

On Sunday, 25 October, with our aquanauts back ashore, no one in the habitat, and no surface divers in the water, I spent the day in conference with some of my dissident aquanauts; with Morgan Wells, our safety director; and later with my great old friend Walter Feinberg, who drove up from Boston for a welcome visit. With none of the customary emergencies, we were free to sit at ease in my corner room, staring out to sea for long periods, and to talk at length of people, programs, and progress that we had witnessed together or apart over seventeen years. Somehow, though, I sensed that something was missing. Within three-quarters of an hour, my finger was square on the button.

"Walter, where in hell is your pipe?" There had to be a story here, and I awaited a great one.

"I quit two years ago, George."

Just like that. Walter Feinberg, scuba buff, world traveler, but always Walter with a fresh corncob jutting from the classic features; always Feinberg with an attaché case full of Missouri meerschaums; always Walter F., the urbane Bostonian, ever spreading the gospel of the pipe. But no longer. It was my same great friend, but his countenance had changed, as if he had suffered an esoteric conversion. That, of course, would not do at all. I hastened to offer a spare briar, but he fended me off gently, then asked in a quiet tone, "Why are you still smoking a pipe?"

Why indeed? I paused to empty and refill my Italian import, then establish a fire therein. Why indeed? I commenced my little lecture in the least patronizing fashion at my command.

"Walter, I ask you to think carefully about what I'm about to say. First I ask you to ponder on why any man should smoke a pipe. Assuredly, it is not for pleasure. Why would any thinking man take up a lifelong servitude to a foreign object that chips, distorts, and irrevocably stains his teeth? Why a permanent union with the pipe, which keeps the tongue continuously raw, produces a halitosis equaled only by cheap moonshine liquor, requires more maintenance than a vintage Edsel, and consistently burns holes in any garment worn?"

At this point, Walter made a feeble gesture of protest, but I buried him with another sweep of oratory.

"Why does an otherwise rational man elect to smoke a pipe, thereby keeping his marital status in delicate jeopardy? Have you ever wondered about these things? Of course you have, Walter, and that may be why you threw in the towel two years ago." If there was a note of disdain in my voice, I sought to conceal it. But there was more to say.

"No, Walter. A man smokes a pipe for one single reason: it is a perfect intellectual security blanket. Sometimes an inconsiderate person will ask you a question that calls for an intelligent answer. Your mind having been adrift, your jaw drops open, and you feel yourself on the verge of stupid babbling. So what is your reflex response? You jam that stinking pipe in the gaping orifice, suck deeply upon it, and raise your eyebrows. And that, Walter, gets you off the hook and your friend will look for his answer elsewhere."

Walter Feinberg gazed at me for long moments, as if assessing his chances of unhindered escape. Then, cautiously, he began to discuss Boston politics. Just as I warmed to this fascinating subject, he clasped my hand in fond farewell. For a while I thought perhaps Walter hadn't got my message, but later it occurred to me that Walter Feinberg had matured beyond the need of an intellectual crutch such as a pipe.

At the beginning of November, the last month of my naval career, I was 2,400 miles from home with nowhere to hide until the job was spelled out. That was the way I had started in the Navy and I had never regretted a day of it. A wise man once said that the beginning and the end are inevitably the same.

The weather had again turned against us, but at least we had no people in the *Helgoland* and no surface support divers at sea. Wind was steadily out of the northeast, and what I had hoped to be a passing squall had become a condition that thereabouts was not looked on favorably. New Englanders are rarely wrong about the weather, and when I called John McLaughlin after midnight, expressing concern about the rising wind, his most positive comment was to the effect that our past four days of calm had been 48 hours too long anyhow. From John, that amounted to a declamatory address, since he limited his remarks usually to three categories: yes, no, and silence. In retrospect, such brevity had a point, since there was little else to be said when it came to New England weather. Our age-old problem was learning to beat the wind and the sea. I had grown old in this game, perhaps too old, but I knew that humans had to fight for every inch they gained from the friendly sea, and much harder when she became our enemy.

On 1 November our plans called for committing our last team to a shallow but still quite deadly depth. Late the night before, I had talked to each diver apart. In the long run, I guess all I could say was that I would never desert them. What that meant to an aquanaut on the bottom, I could not honestly say. Most of the jobs that were to be done on site on 1 November called for expert seamanship amidst our fleet of precious buoys, and expert deck handling of bulky gear, not to mention diving difficulties. Our people would go out regardless, to do their best under bad circumstances. Good sailors are just made that way.

At daybreak, the weather looked like a sure go. Teams assembled, equipment was loaded and offloaded, and our little armada set sail. By midmorning, it had returned to port in the face of rising winds and seas. At that point there was little that we could do in further preparation for the next team's stay. It would be beautiful someday to have surface vessels that withstand wind and wave and still relate to specific bottom sites with accuracy while staying on station.

That evening I prepared to make my first after-dark foray beyond the Emerson Inn. The day before, a group of very old biddies, whose hobby happened to be antique dolls, had arrived at the inn. By late afternoon, the ranks of these ancient collectors had swelled to a veritable mob that produced a growing cacophony as the beautiful collections of dolls were displayed. As nearly as I could tell from my balcony watch station, a few verbal encounters took place, but even my granddaughters fight over their dolls.

The upshot of this invasion was that the Emerson Inn management offered the lot of us scruffy sailors a night on the town, at management expense, to relieve congestion and spare the kitchen help. We accepted without a murmur, and a gaggle of German and American citizens took off for Mama Felice's Italian restaurant in South Boston. An old friend of Mama's had made prior arrangements for us, so, on arrival, we shouldered our way through the muttering mob waiting in line, straight back to a reserved room. The tables were already laden with good wines and trays of hors d'oeuvres. Soon thereafter, the serious eating began.

First, great bowls of delicious fettucini, followed by three seafood courses as a warmup. There were two meat courses, chicken and veal, with side dishes of ham and cheeses. Then we were served salad with another dish I couldn't identify, topped off by a bowl of marinated mushrooms. When the pastry course finally arrived, even my appetite was satiated and I lacked nerve to ask for a doggie bag. This is a true account of the feast, the likes of which I never saw in Italy. What the bill was, I'll never know, as the Emerson Inn, true to its word, picked up the tab. We were all home and abed before midnight, while below us the doll collectors were just shifting into second gear. On a nearby hill, the late Ralph Waldo Emerson probably turned fitfully in his weathered grave.

Although at the time I may have thought of 1 November as a day of no operational significance, I would qualify that description to a day of little visible significance. A delicate, massive undersea habitat lay offshore, unattended and unmonitored. While we sat ashore, the seas might have been doing small but steady bits of devastation. We had no way of knowing, so we tended to assume that all was as we left it, though history tells us this is not likely. Any unattended system placed in the ocean will degrade at an accelerated rate, and something as complex as the *Helgoland* habitat would be particularly vulnerable. Considering the state of the art of remote sensing, it was disturbing to know that not one iota of information about our vital components could be received ashore unless a diver was in situ to relay it. I strongly recommended that the situation be remedied.

I wakened on 2 November to a period of flat calm and no wind, and it remained so all day. However, the unattended and unmonitored habitat complex had suffered from the force of the weather and perhaps from internal complications. During our enforced absence from the site, the high-pressure line to the habitat reservoirs sprang a leak, and a weak link in one of the energy buoy three-way anchor chains parted. The status of the batteries in the habitat could not be ascertained. There was a possible problem with the PTCs, but the radio communication, being garbled, did not clarify what the problem was. I waited for the verbal report from all hands who had been at the scene that day.

This much I knew: every diver who went out there burned up his daily bottom time just staying ahead of the game. They worked like beavers, and at considerable hazard, playing catch-up, which was no fun in any man's book. We were promised another turn of good weather but I decided to wait and see, feeling a south wind at my back that had not been there before. But our crews arrived back ashore with no casualties, and for that I was grateful every time. It was a blessing to know that they were professionals who were equal to the task.

I suppose that a matter of deepest concern to me during the FISSH project was the basic philosophy of my German colleagues, whose friendship I cherished and whose professional capacities were beyond question. Nonetheless, they had a loyalty to their diving complex that sometimes transcended common sense. In my experience, no diver habitat system was immune to honest criticism; certainly, there were obvious flaws in the *Helgoland* complex, as there had been in all our Sealab engineering efforts. According to the Germans' apparent logic, however, everything that had functioned perfectly in the past would not fail thereafter; there could be no predictable problems because the engineering was perfect; when failures became apparent, blame the New England weather, intrusive fishermen, or lack of preparation on the Americans' side of the Atlantic. Such arguments were not always easy to rebut; in general, it seemed best not to rebut them at all, since all of us were deep in this exercise together. Yet there had to be a reckoning one day, when our colleagues would benefit from our value judgments. No one among us had total and ultimate wisdom, but I knew the American team could contribute a great deal to the future operations of *Helgoland*, wherever they might be accomplished.

Our divers came back with a parted compressed-air hose, chafed by wash action against a point of metal until it split, and other, less formidable problems, all generated by our adversary the sea. At best, it seemed that we could not put aquanauts on the bottom for several more days. That would chop us to about two weeks of active bottom time remaining, assuming good weather throughout, which I dared not take for granted. The problem, however, was not even that simple. The positioning and coordinated monitoring of the hydroacoustic array called for relatively calm seas, and those we were not likely to have.

After weeks of silence on the matter of physiological effects during the FISSH operation, I found time on 2 November to give an interim report. Earlier in the project, Tony and I had agreed to obtain one baseline and at least two postdive hematologic scans on each of our saturation divers. As we scanned the results, we were startled to see at first glance that no statistically significant drop in red blood cells had occurred; hematocrits had not dropped perceptibly; and there was no measurable decrease in vital capacity.

To me, the readings constituted a real mystery. First, to set the results in a frame of reference: we had exposed men at atmospheric pressures equivalent to 32–34 meters of salt water for periods of up to eighteen days. The pO_2 ranged from 0.35 ATA to 0.80 ATA, with an estimated average of nearly 0.60 ATA throughout the exposure. The rest of the decompression story has been told, so it will not be repeated. Suffice it to say that our FISSH subjects had been exposed to inordinate oxygen partial pressures, coupled with increased gas densities, yet had shown no discernible hematologic or pulmonary changes.

Our findings were so obverse to any previously published values that Tony and I were at a complete loss as to what to do about ultimate publication. I could not question the reliability of our hospital laboratory, which was above reproach. To this end, I had taken the precaution of meeting with Dr. Cornetta, pathologist in charge, to guarantee the accuracy of the reports by assuring that the same technologists would make the determinations each day, and even that the Coulter counter would not be used for platelet counts. Nonetheless, the hematologic and pulmonary readings were totally normal.

Was it possible that, working in a gray zone between normalcy and evident toxicity, we had found a new and satisfactory milieu for human survival? Common sense told me otherwise. Was it possible that during the prolonged period between the start of decompression and the time of the next blood sampling—85 hours on average—a physiological restitution had taken place? Such was not the case in any of my HeO_2 saturation dives, where recovery took up to three weeks. Beyond the hematologic question, why was there no evidence of impairment of pulmonary function? The answers to the provocative questions raised were beyond my immediate understanding.

In the conduct of the FISSH operation, meanwhile, all went as if especially ordained. The sea was millpond in character, and no fewer than sixteen divers set out on their assigned missions. As radio reports were very encouraging, on 4 November I set about a little experiment planned some days previously. The equipment had finally been put in order and four volunteer subjects were available. Accordingly, Operation Pantyhose got underway.

Since the one-time occasion when we had surfaced the

PTC without loss of human cargo, a few of us had expressed deep concern about the reliability of the CO_2 scrubber system of the capsule. Looking ahead to the ever-present possibility that the aquanauts in a loaded PTC might have to survive as long as 8 hours without ventilation, we cast about for a passive system of CO_2 scrubbing, quite independent of electrical power. Clearly, random scattering of a CO_2-absorbent material within the PTC was untenable, and individual closed-circuit breathing units seemed inadvisable. But how about simply filling a few pairs of ladies' pantyhose with absorbent, and hanging them in the capsule? The idea had appeal, despite its naïveté, so we launched the project, using two pairs of pantyhose, one black, one red, filled with a total of 8.6 kilos of commercial absorbent.

For the actual experiment, we locked four volunteers, one of them female, in the inner lock of the Draeger chamber (which had a 3,000-liter volume), supplied them with an O_2 monitor and a batch of CO_2 sniffer tubes, and left it up to the pantyhose array to do its bit. To supply metabolic oxygen requirements, I maintained a constant flow of 2.5 liters of oxygen per minute, which kept the chamber atmosphere at a perfect 21 percent throughout the procedure. Both CO_2 and O_2 levels were determined inside the chamber at 15-minute intervals and recorded outside, while I maintained constant visual and voice contact with our subjects.

As you might guess, Morgan Wells and I were a bit edgy at first because the CO_2 levels in the inner lock could be expected to rise by 0.82 percent every 15 minutes, a rate that left little leeway. Still, we had plenty of safeguards so we started the show.

Both Morgan and I were a little stunned when the first 15-minute CO_2 reading came out a fat 1.5 percent, and rose quickly thereafter to 2.25 percent. We had faith in the system, however, and stuck to our guns. As chamber humidity commenced to rise, the galloping CO_2 slope simmered down, and after almost 3 hours the level stayed steady between 2.75 and 3.0 percent. By this time, we had already designed the MK II model pantyhose scrubber, capable of 75 percent greater efficiency. We released our volunteers none the worse for the experience. We then made up the MK II pantyhose scrubbers, sealed them in plastic bags, and installed them in the PTC. I found this kind of improvised and on-scene experimental work fascinating; in this case, it instilled a sense of confidence in the aquanauts as well.

On 5 November I arose before daybreak and, after a cup of tea, wandered to the lower porch to sample the wind and assess the current sea state. There was no wind, and the sea was flat calm. As I stood enjoying the colors of the rising sun, I was joined by John McLaughlin and Dick Cooper, both early risers like me. I remarked that we would be smart to start operations early in the day, to take advantage of the calm hours before the sea and wind began to make up. Since our experience had been that normal operations usually became impossible shortly after midday, we felt that getting on site by 0700 might make the difference between a day's effort and an abort. We resolved to raise the question at our meeting that evening.

About 0930 the diving crew assembled and left for their assigned dive vessels. At 1115, radio informed us that a dive boat was on station, and divers were preparing to enter the water for the first and most difficult job of the day—connecting the topside portion of the *Helgoland*'s umbilical just beneath the surging energy buoy, a monstrous pile of floating hardware. By then the seas had built to 5 feet, not formidable but enough to cut a diver apart should he be caught in any bight beneath the heaving mass 3 meters above—else beat his brains out on a single down-surge. Our divers persisted, but without success.

At 1400 hours came word that all diver operations had been shut down, and that the dive boat *Atlantic Twin* was returning to port, marking another day of operational failure. I went to our evening directors' meeting in a glum mood to start with, and my spirits hit rock bottom when I heard that, should the surface support divers make the below-buoy connection tomorrow, two members of our saturation team would go to the bottom to make the vital connection with the habitat. Making the topside connection might take up to 3 hours to complete, they affirmed, and they might even fail at that. Such being the case, the habitat would not be ready for occupancy, and aquanauts Cliff Newell and Johan Rediske would then have to enter the De-KomRaum for decompression, recompression, and return to the surface.

I protested against the plan as outlined. If successful in making the bottom umbilical connection, I proposed, the divers would simply enter the habitat as the first two members of the team. Mind you, these were two of the aquanauts who had defected some days past, but they had then come back after a talk with me. They were willing to put their lives on the line and trust in my ability to pull the rabbit out of the hat. Facing the problem clearly, I told them my dilemma. They understood that I would be unhappy to violate once more every sacrosanct rule in the book; still, the undersea connection to the habitat had to be made, and I had every reason to believe that Cliff and Johan could make it, enter the habitat, and take up their jobs as saturation divers on our final team. If they failed to make the connection, the decompression problem would be mine for solution. In the general interest of the FISSH program, we agreed to do it.

Other matters went well enough during this session until longtime aquanaut Dick Cooper raised his calm voice. Dick said it seemed to him that we had been missing the advantage that the early morning flat sea conditions would give us in accomplishing our critical jobs before midday, when all meteorological hell tended to break loose. It seemed only logical to him that we should arise around 0400 and get out while the sea was calm to get the job done.

To me, this sounded reasonable. Since the beginning of this exercise, I had never heard a complaint about long hours from any member of the American or German crew, though I had detected some reluctance on the part of a few German crew members to stand chamber watches when we were doing operational tests. Yet I had great confidence in round-the-clock support by the German crew. Imagine my surprise, then, when Klaus Oldag shouted that no one on his team was to be awakened before 0600 hours. We were challenged, once more, in a game of one-upmanship. It would have been easy to say, "Very well, we'll take our American divers out to do the job while the Germans get their sleep." But the first job on the agenda was to connect the umbilical to the energy station buoy, and that equipment was verboten to all Americans. From our side of the house, Ian Ellis simply announced that he would load his gear before retiring, and all the other American divers agreed. The impression was made that our crew would be ready, no matter how long it took the Germans to wake up. We broke for the night, barely in time to avoid words.

But all's well that ends well. Cliff and Johan got into the habitat on 6 November, found all was well, hooked up the terminal of the umbilical, and settled in as resident aquanauts. Considering the rigorous nature of every dive, I marveled that we had not yet had a discernible case of bends. Even more amazing, we hadn't seen a traumatic accident, or worse, in the course of daily shipboard work, despite handling many fathoms of chain, line, and wire from the decks of pitching, yawing, rolling vessels. There had been close calls that I knew about and doubtless scores of others that had never come to my attention. Now my last vigil began.

There were aquanauts on the bottom, and my sole purpose was to keep them alive and healthy and, finally, to surface them in good shape. That I dedicated myself to do.

After Wes and Ian got down safely to join Cliff and Johan, they immediately made exploratory dives for bottom re-orientation, then freshly memorized the systems of the habitat. That was the correct priority; they would have time enough to sleep later. It was most important first to scout the backyard, then learn again the interior of the den. Such is the way of a cat when dropped in a strange neighborhood, and it is often a lifesaving habit. Even a city dog turns around a few times before bedding down.

Later we began to dilute the atmosphere in the habitat by pumping in nitrogen. Although we discharged eleven bottles, the pO_2 level came down to only 0.42 ATA, and I suspected a compressed-air leak. Later the pO_2 edged down to the vicinity of 0.39, meaning that a few more bottles of nitrogen should make it about right, in the 0.35 range, unless we developed another compressed-air leak, which would be a shame.

I didn't sleep after the early morning check on 8 November as the damned foghorns kept up their melancholy monotone, beginning at 0200. The fog was amply confirmed by my eyes. We were socked in, and for how long was anybody's guess. It was a spooky business, since the energy buoy sat naked 11 miles out at sea, a vulnerable obstacle for coastal traffic in the fog. Loss of that energy buoy would throw us into an emergency mode in short order. I finally went below for a short walk and a long breakfast. As I arose from the table, Tony approached with a German geologist–diver in tow.

The diver, Galib Demiray, had made a single descent to the habitat on the previous day. He was vague about his bottom time, but our radio log showed 19 minutes for him, surface to surface. At that tide, the habitat depth was 34.5 meters. Further questioning revealed that he had come up in the exit hatch for conversation and coffee with the aquanauts. *Das*, of course, *ist verboten!* Demiray had begun to have pain in his left leg and hip pain at about midnight—5 hours after surfacing—but had waited until breakfast time to report it, although he had known that I had been up and available most of the night. Well, that was water over the dam, and after Tony assured me that Demiray's behavior did not indicate a neurological deficit but was simply a case of pain-only bends, we took him straight to the decompression chamber for a test of pressure and treatment.

Demiray's case of bends could not be considered routine, since it exhibited a delayed onset of 5 hours in symptoms and an added delay in treatment. Given those circumstances alone, I would probably have elected to start treatment at table V, shifting to the longer oxygen tables according to the patient's response. But technically, Demiray was not my patient. He was Tony's patient, and Tony was committed to the treatment tables spelled out in a document endorsed by the Federal Republic of Germany. Under those circumstances, it did not behoove me to recommend any other specific mode of treatment; the program had been well written, in Tony's eyes, and I would abide by it. We thus began a bends treatment at table III, most despised by men of experience in undersea medicine. But it was not my nickel in the machine and, for a change, I kept my mouth shut. All went well, and Demiray was free of all pain at 165 feet, or 50 meters. Though table III had failed me often in the past, his was a pain-only bend and, although treatment was delayed, we probably overtreated.

I was sorry to see Demiray hit, since our track record had been so good. The rules of chance play no favorites, and if we calculated otherwise, the efficiency of all undersea operations would become as nothing. In all events, Demiray was comfortable at 10 meters, where he would continue a long

10-hour soak, and then surface some 4 hours later. I reckoned the event had a sobering influence on all the divers.

As we talked bilingually in our evening meeting, one man said to me, "How can this possibly have happened to Demiray, who was my dive partner? He got bends but I did not, on the identical dive."

I could only give him my stock answer: "Should you and Almighty God make the same dive together, God might be bent while you were spared. Perhaps God would know how such a thing can happen, but it goes beyond my wisdom."

For my part, I had never thought to terminate my Navy career sitting beside a German chamber, treating a Turkish diver for bends in a canvas tent that threatened imminent collapse in a New England gale. But Demiray's treatment went remarkably well. While I slept for an hour after supper, Tony locked in and did a careful examination of our patient, who by that time was proclaiming that the entire painful episode was simply an acute exacerbation of the rheumatism contracted last year in northern Germany. As I told Tony, this reaction helped prove my contention that each case of the bends is simply an act of God—if one is to believe the victim. Tony and I agreed that we should patent the method of treatment as an instant cure for crippling arthritis. Meanwhile, we would go on calling it bends and blaming it largely on bad diving habits.

As we neared the final phase of Demiray's treatment, I realized that in all likelihood it would be my last professional supervision of a bends case. I did not regret the passing of this milestone. If memory served me right, of the six hundred or more cases of bends I had treated, fewer than a score of the victims had ever said as much as a simple "thank you." I believed there was an excellent psychological reason for this lapse in common courtesy, which reflected no credit on basic human behavior. Given a successful appendectomy, tonsillectomy, or ablation of hemorrhoids, physicians usually receive a word of thanks from the grateful patient. Bends victims, however, unlike the other patients named, have usually brought their houses down about their heads through their own negligence. Those who are stricken will seek any path of explanation that does not expose their own mistakes. As Bob Workman and I had come to know, all too often such accident victims, even if successfully treated, would turn some years later to a contingency-fee attorney and bring all parties to court to account for the onset of arthritis, hearing loss, impotency, or other frailty that might have developed over the years. Quite possibly, alcoholic cirrhosis of the liver is the only ailment not yet attributed to treatment of a bends case.

Obviously, the business of diving had never been rated a safe occupation, and for that reason we were all paid more for our work than might otherwise have been the case. The hazards were clear to all divers. Nonetheless, some elected to take the risk, and as a natural result, some of them died and some who survived did so only with physical disabilities. I deplored it, and did what little I could to alleviate the effects, but the choice was theirs, just as it was my own decision to become a diver. In this world, each one must make personal choices and then accept the consequences.

Meanwhile, a south wind blew away the fog and brought a rise in temperature that made the watch more agreeable. Only 3 hours of the watch remained, and I knew I could do that standing on my head. The decompression treatment had gone rapidly, thanks to my German friends who kept me company, who were always in good spirits, and who tolerated my poor grammar in rare good taste. My last companion of the night was Hans Belau, and better company I couldn't ask for. The man had a rare combination of humor and philosophy and was a remarkably good diver to boot. When we brought Demiray up to 2 meters of depth, he was in splendid health and spirits.

It hardly seemed possible that in less than a fortnight FISSH would end. Despite our difficulties, caused largely by weather, we had accomplished a great deal, although scientific achievements do not usually surface for several years after the event. Personally, I had made a slew of lifetime friends and learned a hell of a lot, albeit so late in life. The mission had renewed my flagging faith in man's earnest desire for undersea exploration and exploitation, and I found comfort in the thought.

15

FISSH
(November 1975)

On 13 November the moor was laid for the *Atlantic Twin*. Perhaps it is easy enough to lay the legs of a moor in the empty ocean, but to do the same amidst a cluster of buoys and almost squarely atop the habitat called for uncommon degrees of seamanship and ship handling. The uncanny capacity of some sailors to feel as one with the current, the wind, and the bones of their ship is an unending source of wonder to me. It cannot be a matter of book learning, as I've seen it exhibited by seamen who could scarcely read, much less comprehend trigonometry. It may be an art achieved by collective use of the senses, without regard to instruments. I have seen such sensitivity in medical practitioners who could make an accurate diagnosis by touch of the hand, smell of the body, and prolonged visual observation of the patient. I pray that this instinct will never be lost to mankind.

Early that morning, Tony took Hans Belau to the hospital for a few days' observation and intensive treatment of the duodenal ulcer that had showed up in his X-ray series. That casualty was another we could ill afford, but we had no choice in the matter. Hans was to be flown back to Germany, separating him from the stress of this operation. Manfried Pose replaced him as senior diver. One at a time, we were losing our best divers, to a minor illness here, a major problem there. From the initial count of twenty-two, we now had fourteen, about eight of whom I rated as top-notch.

The weather gave us a break later on 13 November when wind veered to the south as predicted; as its force was only around 5 knots, all our crew went to sea. The Soviets were afloat as well, but they would not dive, being aboard only to learn our mode of operation. Had their observations about our procedures been accurate, they might have set the Soviet Navy back by at least ten years. Perhaps the Soviets deserved no better. Had they sent us divers rather than topside observers, my outlook might have been different. Besides, it was a bit disturbing each morning to see the hammer and sickle raised to the yardarm, although it had every right to be there.

Early on 14 November, Ian called me from the habitat to report a severe headache and request translation of the titles in our medical kit. I responded with, "*Hast du, mein lieber freund, ein* noggin gesplitter, *aber* merelich *ein* simplischer noggin-knocker?*" The German language kind of got to me after a while, and from then on I spoke a jargon understood only by *Helgoland* aquanauts. The remarkable result was that Ian took in my message in its entirety.

I asked when they had last changed their CO_2 scrubber charges. That chore turned out to have been done as long as 36 hours previously. The oversight was immediately corrected from the habitat's stock of Sodasorb, which was large enough to last through New Year's. Meanwhile, I told Ian to take two pills from package No. 2. I knew that those German tablets were intended for rheumatism, but anything that eased the pain of rheumatism would surely cure a headache. Despite the haphazard therapy, Ian Ellis reported loss of his "noggin gesplitter" half an hour later.

Our council that evening took up the problem of increasing vulgarity in the speech of our aquanauts. On the West Coast it had mattered little, but in Massachusetts we were plagued with busybodies who had receptive antennae that picked up every word emitted from the habitat. Needless to say, some of the discourse had in fact been salacious, quite unfit for the general public. However, I considered this liberty of language to be simply part of the aquanaut break-away phenomenon that must be anticipated.

The state of the weather might have contributed to my restless state on 14 November. From Central Control I could read the seas and the wind all too well, and I tried to translate them into the projected activities of the day ahead. A troubled sea state was of little concern when it came to diving performance itself. Once divers were 3 meters down, they were quite safe, and though they might be bumped about on the surface return, they were not likely to suffer real injury.

Such was not the case with surface support sailors, who had to handle heavy gear for overboard deployment under adverse sea conditions. One single lapse of thought, one foot caught in a bight of line, and a serious casualty was guaranteed. Our aquanauts deserved the continuous protection we gave them, but sometimes we forgot the sailors afloat, a mistake on our part. Whatever the limits of humans, both topside and below, there was a breaking point, and we were close to it.

A few weeks previously, I had committed a minor tactical error in revealing to a member of our group that I had a birthday in November (14 November, to be exact). At the time, I thought nothing of that revelation, since at my advanced years such milestones should have had little significance. Not so for my aquanauts and the staff of the Ralph Waldo Emerson Inn. I had sorely underestimated the cunning of this fabulous group, for they had plotted and planned the damnedest birthday party of the century. It started after a supper of fresh halibut, when Tony and I lingered over coffee to talk of his student years in Paris (as it turned out, he had lived a scant few blocks from my former garret at 5 Rue du Colisee, near the Arc de Triomphe). As we talked, the gracious Emerson Inn waitresses brought me gifts, sang "Happy Birthday," and kissed me through my tangle of beard. I was quite moved by the simple ceremony and hoped I indicated as much.

About an hour later, as I lay abed reading *Science* and listening to good radio music, came a gentle knock on the door. It was Tony, asking if I would step down to the basement conference room. Fearing bad news, I slipped on a pair of sneakers and quickly went below. In the recreation room I found forty people or more, an enormous cake, and presents of every description. Within minutes, I was regaled with "Happy Birthday" greetings in Polish, German, English, and finally in Russian by our recently arrived Soviet scientists. The cake had the requisite sixty candles, which I managed to blow out in a single breath.

After the presentation of extraordinary gifts from every nation, there followed much kissing, hugging, and shaking of the hand. A large bottle of Polish vodka appeared, soon to be joined by several more, not to mention a few gallons of German beer. It was then that one of our Soviet scientists found the old upright piano and someone dug up a guitar. At that point, the affair became quite lively, with dancing and kissing and not a little tomfoolery.

After a bit, I sneaked away to Central Control to describe the event to my aquanauts down in *Helgoland*. Topside, my radio reception was perfect, indeed a bit too clear for public ears. It started with a ribald poem that Johann Rediske had composed for the occasion, which wouldn't have been publishable in *Penthouse*, or even *Outhouse*. Next came no end of friendly comments from my narcotized aquanauts covering a narrow range of interests having to do with sex, and I bantered some indelicate subjects back and forth, as is my habit. At that point I received a telephone call from our local harbormaster, who reminded me that channel 22 was being monitored by half the female population of Gloucester. No complaints had been registered, however. When at last I crept to my room, I was accompanied by loud songs and much opening and closing of doors. I would venture to say that the morals of my puritanical Yankee brethren would no more stand up under scrutiny than those of my warm-blooded southern kin. Kipling probably said it best: "For the Colonel's Lady and Judy O'Grady / Are sisters under the skin." (Of course, the social implications of those lines probably cost Kipling the title poet laureate, which was awarded to John Masefield instead.)

As an occupant of the Emerson Inn's upper deck, I was acutely conscious of the threat of sudden conflagration in the hostelry, since the ancient structure was a veritable tinderbox, although the average room temperature would surely have defied spontaneous combustion. In compliance with recent local regulations, the hotel had been fitted out with a new fire detection system. Early on 16 November, the fire alarm brought us all together on the front lawn in varying states of undress. Although it turned out to be a lively gathering and I understood the need for unscheduled fire drills, I would rather have stayed abed. After that, the alarm tended to sound almost hourly, and we became hardened to impending disaster at the inn, stowing our valuables in the FISSH decompression tent, to be sorted out later. Regardless of the fire probability, the inn was a delightful place to stay, managed and assisted by the warmest people in my experience. Somehow they came to love our bunch of divers and mechanics, and we returned their affection with all our hearts.

Although snow and high winds were predicted for that day, neither transpired. However, our two remaining surface support vessels both broke down, so we accomplished very little. We were a pretty tired crew and the day of rest was welcomed. I used the breather to review the overall situation. Sometimes retrospective review improves the odds of success next time, although the record is seldom read or heeded. I would say that the scope of the FISSH project was perhaps too great, considering the men, money, and equipment available. Inflated expectations are not uncommon, but they are especially serious when the project at hand is international in character. Somehow, we never anticipate that men and machinery wear out all too rapidly under adverse circumstances; that about 20 percent of total expenditures are both unforeseeable and absolutely necessary; and that any operation to be conducted in open sea will lose at least one day out of five to adverse weather.

Those are simple facts of life, well known by operators, but not at all understood by planners. Possibly the simple geographical separation and lack of communications between the two groups was a major factor in the FISSH project. I believe, however, that a refusal to read and heed past history is fearfully significant. All of us who have worked in the field know that an underwater job will inevitably take three times longer than the same task ashore, but planners never take this factor into account. As a result, divers are driven to outer physical limits and beyond.

The FISSH project was an exercise that combined professional personalities of five nations. The language barrier was not a serious obstacle, largely because our foreign friends became remarkably fluent in our tongue, even though we did not meet them halfway in this regard. Perhaps more alienating than language differences were the differences in philosophical attitudes, too often many degrees apart. It was a point of national pride with the Germans that mechanical failures were virtually impossible in any of their systems. When such failures inevitably occurred, they tended to attribute the breakdowns to forces beyond their foresight or control, an understandable pattern of rationalization.

Our philosophy was different but no less vulnerable. American design engineers also considered their work immune to failure, but we users recognized the fallibility of both the engineers and the fabricators' quality control. The end result was different, in that the German user never expected nor was prepared for the hardware to fail, whereas we took along perhaps an excess of spare parts.

Beyond such considerations, a grossly underestimated factor was the progressive emotional and physical fatigue inevitable in such an effort. The emotional toll became evident in personality clashes that, barely evident early on, assumed ridiculous and dangerous proportions as the days and nights passed. Topside personnel became polarized even as the bottom-dwellers felt glued together by common purpose and constant danger. Ideally, a project leader should identify potential personality clashes early on and take immediate steps to isolate one of the two antagonists from the operation. In the Sealab projects, able participants, officers and enlisted alike, had to be banished from the scene, to the credit of no one involved. The same phenomenon occurred in the FISSH program.

Like emotional fatigue, the physically depleting effect of relentless round-the-clock operations was too seldom taken into account. The concept of the diver as a tireless man of iron was ridiculous. Reality demanded that we either shorten the time of the mission, else soften the tasks imposed on the crew in deep-sea diving.

The FISSH operation wound down rapidly. The hydroacoustic sonification experiment terminated with a sad track record, perhaps doomed as a project that went straight from laboratory prototype to a massive open-sea effort. Still, I felt deeply for the dedicated people who had labored so hard to make it work against such odds.

That being the final phase of the program, we began decompression of our last team on 17 November 1975. Conditions seemed perfect: clear weather, calm seas, and a smooth button-up of the *Helgoland* habitat. Tony and I smiled confidently as the aquanauts informed us they were entering the

DeKomRaum to begin decompression. Surely, this one time, things would go as planned, and Chris Lambertsen's table would be used in its pure and unabridged version. Two minutes later, word came from below that power had failed; the diesels in the energy buoy could not be started. Unhappily, we called our aquanauts back to the main room and settled back for another delay. By a rare stroke of luck, the problem was diagnosed and quickly repaired.

It was still smooth sailing twenty-four hours later, when Claude Harvey came in to relieve me and I packed my gear for an early morning departure for Panama City. I delivered a touching speech of farewell to my aquanauts: "So long, you guys." The Bronx cheers in return warmed my heart.

To avoid another long round of emotional farewells at the Emerson Inn, I posted a notice on the bulletin board to the effect that I had been kidnaped by a wandering group of gypsies. Anything to avoid a tearful parting, especially when I'm the one who cries.

To FISSH, *Helgoland*, and all the gang ashore and afloat, Pappy says, "auf Wiedersehen."

Epilogue

Another brief message from your obedient servant. I informed you previously that a few of my old shipmates and cronies from Genesis and Sealab days were planning a brief afternoon roundelay with hearty good wishes and perhaps a pint of brew to mark my retirement from the Navy on 1 December 1975. The revised program called for a timber-rattling series of Decemberfests cum orgy to commence on the afternoon of Saturday, 29 November, continuing until the late hours of 1 December, with a brief interlude to observe my formal departure from the Navy. For this purpose, my mates rented out every structure left standing on the beach in the wake of the hurricane.

I could not stem this tide of warm fellowship but I gave to all upright citizens a last chance to stand clear.

Affectionately,
Pappy

Index

About the Author

George Foote Bond, Captain (MC) USN, retired from the U.S. Navy on 1 December 1975. As a founding member of the Institute of Diving in Panama City, Florida, he served it as president emeritus and edited a newsletter for deep-sea divers until his death on 3 January 1983. Captain Bond rests alongside his wife, Marjorie, on a bluff overlooking the Church of the Transfiguration in Hickory Nut Gorge, Bat Cave, North Carolina.

On 17 May 1991, the Ocean Simulation Facility at the Naval Coastal Systems Laboratory, Panama City Beach, Florida, which Captain Bond had praised as "the finest hyperbaric complex I shall see in my lifetime," honored him by emblazoning his name above its entrance. The Institute of Diving and the Museum of Man in the Sea, so generous in providing access to the chronicles and personal correspondence of Captain Bond, is open to the public year-round at 17314 Back Beach Road in Panama City Beach. A prominent feature of the museum's collection of historical diving apparatus is the restored habitat Sealab I.

About the Editor

After receiving the Bachelor of Education degree from Rhode Island College, Helen Siiteri alternated teaching in an elementary grade school and acting in New England summer stock companies. She also served as a storyteller in the New York Public Library, and, following marriage to a research biochemist, she adapted her favorite children's story to read to their five children. *The Adventures of Nicholas* was published by Scholastic Press and for many years was featured in their paperback book club.

When the family moved to Northern California, Ms. Siiteri was hired by an ocean engineering firm to research and prepare ship salvage reports and a history of ship salvage in the United States Navy. *Papa Topside* grew out of that research.

She is currently working on a novel based on the grounding of the Maltese freighter *Eldia* on Nauset Beach, Cape Cod, in 1984.

THE NAVAL INSTITUTE PRESS

PAPA TOPSIDE
The Sealab Chronicles of Capt. George F. Bond, USN

Designed by Pamela Lewis Schnitter

Set in Simoncini Garamond
by Brushwood Graphics, Inc.
Baltimore, Maryland

Printed on 50-lb. Glatfelter smooth antique white
and bound in Holliston Roxite B linen
by The Maple-Vail Book Manufacturing Group
York, Pennsylvania